27.50

D1472926

Fundamental Research in
Homogeneous
Catalysis

Fundamental Research in
Homogeneous Catalysis

Edited by

Minoru Tsutsui
Texas A & M University, College Station

and

Renato Ugo
University of Milan, Italy

Plenum Press · New York and London

Library of Congress Cataloging in Publication Data

International Workshop on Fundamental Research in Homogeneous Catalysis, 1st,
  Santa Flavia, Italy, 1976.
  Fundamental research in homogeneous catalysis.

  Includes bibliographies and index.
  1. Catalysis – Congresses. I. Tsutsui, Minoru, 1918-   II. Ugo, Renato.
III. Title.
QD505.I6 1976   541'.395   77-13024
ISBN 0-306-34441-6

QD
505
.I6
1976

Proceedings of the First International Workshop on
Fundamental Research in Homogeneous Catalysis
held at Santa Flavia, Italy, December, 1976

© 1977 Plenum Press, New York
A Division of Plenum Publishing Corporation
227 West 17th Street, New York, N.Y. 10011

Printed in the United States of America

PREFACE

The objective of this workshop on homogeneous catalysis was to identify opportunities for the solution of energy problems and industrial production problems by homogeneous catalysis. The first day of the workshop was devoted to plenary lectures on frontier areas in homogeneous catalysis which set the tone for the workshop. On succeeding days of the workshop, the participants were divided into five working groups for discussion of various aspects of homogeneous catalysis.

Each of the five workshops engaged in extensive discussions and then formulated a rough draft of their report and recommendations. The reports of the working groups were presented at a plenary session and suggestions for changes and revisions were made. These minor revisions were incorporated into the working group report by the Co-Chairmen of the working groups.

This workshop on homogeneous catalysis was sponsored by the National Science Foundation (United States) and by the National Research Council (Italy). Additional financial support was provided by Montedison, E.N.I., and S.I.R.

We wish to thank Mr. William M. Tsutsui for typing and assisting in the editorial work. The Robert A. Welch Foundation Grant A-420 partially supported the time spent by M. Tsutsui for the organization of the workshop and the editorial work of the proceedings.

Organizing Committee, December, 1976

C. Casey
G. P. Chiusoli
J. Halpern
M. Tsutsui, Co-Editor
R. Ugo, Co-Editor

CONTENTS

PART II: WORKING GROUP REPORTS

INTRODUCTION

The diminishing world supply of petroleum has now made the conservation of energy a prime goal of the chemical industry. In the future, additional problems will arise in switching from petroleum to coal as the major source of hydrocarbon raw materials. The major objectives of this workshop were to point out specific areas where fundamental research in homogeneous catalysis will be required to help solve these problems.

Most industrial catalytic processes now employ solid metal surfaces or metal oxides to catalyze the reactions of gaseous or liquid reactants. While heterogeneous catalysis is often the basis for important and efficient processes, heterogeneous catalysts sometimes are far from optimal. Heterogeneous catalysts often provide only limited selectivity to the desired product, which causes waste of chemicals and energy. Moreover, heterogeneous catalysts often require high temperature and consequent waste of energy to achieve rapid reaction rates. For exothermic reactions, problems are often encountered in dissipating the heat of reaction from the solid catalyst surfaces.

The reports of the working groups are very optimistic about the potential uses of homogeneous catalysts in industrial processes. Homogeneous catalysts are discrete molecules containing one or more transition metals and their surrounding ligands. These organometallic catalysts offer a number of potential advantages. Homogeneous catalysts make more efficient use of metals than heterogeneous catalysts since each atom is an active site whereas in heterogeneous catalysis only atoms in a particular site on a metal surface provide active sites. Homogeneous catalysts are potentially more specific since each metal atom is in the same environment; in contrast, the metal atoms on a disordered metal surface experience a variety of environments. Homogeneous catalysts often operate under milder conditions; this is an advantage since there are obvious energy savings and since selectivity for any reaction is increased at lower temperatures. Variations of the ligands on a homogeneous catalyst can lead to systematically improved selectivity. For example, the use of chiral ligand has led to the development of hydrogenation catalysts capable of selectively giving a single enantiomer. Heat dissipation from exothermic reactions run with homogeneous catalysts is not a problem as it is in heterogeneous catalysis. Finally, the mechanisms of homogeneously catalyzed reactions are amenable to study by spectroscopic and kinetic techniques. The increased understanding of homogeneously catalyzed reactions obtained from mechanistic studies can lead to improved control of the reactions and eventually to the rational design of catalytic systems.

The recognition that current petroleum-based hydrocarbon feedstocks must be used more efficiently is another recurrent theme in this report. There is a repeated call for more efficient and more selective catalysts. The eventual shift to hydrocarbon feedstocks derived indirectly from coal is pointed out in the section on carbon monoxide reactions. Moreover, throughout the report there is a recognition that new homogeneous catalysts must be compatible with the impurities in hydrocarbon feedstocks derived from coal.

The limitations of homogeneous catalysts are becoming more clearly recognized. Heterogeneous catalysts offer the possibility of using more than one metal site in catalysis whereas most organometallic compounds have only a single metal atom. Discrete molecular metal cluster compounds are a bridge between homogeneous and heterogeneous catalysis and promise to be important catalysts of the future which combine the advantages of multi-metal systems with those of homogeneous catalysts. Separation of the homogeneous catalyst from reaction mixtures is often difficult. To circumvent this problem, this report considers the role of supported metal complex catalysts and of phase transfer catalysis.

HOMOGENEOUS CATALYTIC ACTIVATION OF OXYGEN FOR SELECTIVE

OXIDATIONS

James E. Lyons

Sun Company Research and Engineering Department

P. O. Box 1135, Marcus Hook, Pennsylvania 19061

INTRODUCTION

Oxygen is a plentiful, inexpensive and highly reactive reagent which, if properly controlled, can effect a variety of useful synthetic transformations. Reactions of hydrocarbons with oxygen are among the least energy-intensive functionalization reactions. Thus, the liquid phase oxidation of organic substrates using metal complexes as catalysts has become a profitable means of obtaining industrially important chemicals.

The majority of liquid phase transition metal catalyzed oxidations of organic compounds fall into three broad categories: (1) free radical autoxidation reactions such as the Mid-Century process for the oxidation of p-xylene to terephthalic acid, eq. 1; (2) reactions involving nucleophilic attack on coordinated

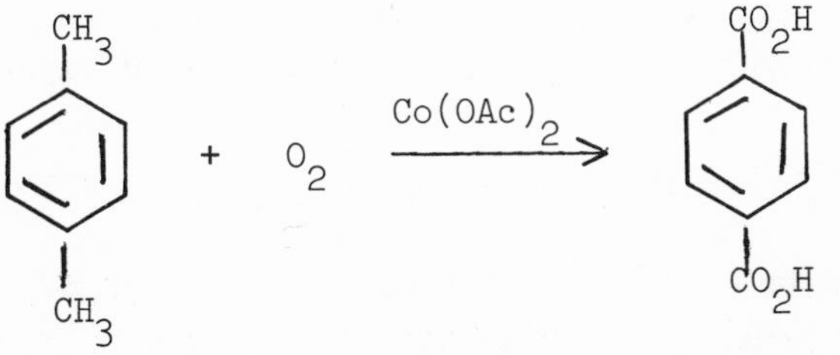

$$\text{(1)}$$

substrates such as the Wacker process for oxidizing ethylene to vinyl acetate, eq. 2; and (3) metal-catalyzed reactions of organic substrates with hydroperoxides such as the reaction of propylene with alkyl hydroperoxides to give propylene oxide in the presence of molybdenum catalysts, eq. 3. Of these three classes of oxidations, only the first represents the actual interaction of dioxygen with an organic substrate.

$$CH_2 = CH_2 + 1/2\ O_2 + HOAc \xrightarrow[CuCl_2]{PdCl_2} CH_2 = CHOAc + H_2O \quad (2)$$

$$CH_3CH = CH_2 + ROOH \xrightarrow{Mo\ compound} CH_3CH\underset{O}{\overset{}{\diagdown\diagup}}CH_2 + ROH \quad (3)$$

Although some autoxidation reactions can be controlled in a useful way, organic substrates more often tend to oxidize in an unselective manner. It is imperative, however, that a reaction be selective if it is to have utility either as an economically attractive process or a convenient laboratory synthesis. Despite the rapid advances made in the area of liquid phase oxidation, much is still to be learned concerning the efficient control of molecular oxygen.

Stimulated by recent success in the catalytic activation of other small molecules, an intensive effort has been made during the past decade to activate molecular oxygen by coordination to a transition metal center. The ultimate goal of this work has been the catalytic transfer of oxygen to a reactive substrate in a selective manner. At this point in time, it is appropriate to review the progress which has been made in this area and attempt to critically assess the status of this work.

Prior to reviewing oxidation reactions carried out in the presence of catalytic quantities of transition metal complexes, it is well to consider the interaction of dioxygen with metal centers and to discuss briefly the reactivity of coordinated oxygen.

DIOXYGEN COMPLEXES

When oxygen is bubbled through organic solutions of transition metal complexes, it is often brought into the coordination sphere of the metal complex. The nature of coordinated dioxygen and consequently its reactivity with the organic substrate is a function of the metal, its oxidation state, its ligand system and reaction conditions. Vaska (1) has represented the ways in which dioxygen is covalently bound to metal centers as shown in Figure 1.

PEROXO COMPLEXES:

side-bonded
O—O = 1.4-1.5A°
diamagnetic

μ-peroxo
O—O = 1.4-1.5A°
diamagnetic

SUPEROXO COMPLEXES:

end-bonded
O—O = 1.25-1.3A°
paramagnetic

μ-superoxo
O—O = 1.3-1.4A°
paramagnetic

Figure 1. Geometries of Peroxo and Superoxo Complexes.

Examples of both peroxo and superoxo complexes have been identified in which oxygen is coordinated to either one metal center or to two metal centers (1-6). Examples of peroxo complexes are far more numerous than are those of transition metal superoxo complexes. In all cases, coordination of dioxygen to the metal center results in lengthening of the O-O bond. This is a consequence of transfer of electron density into π* orbitals of ligated dioxygen. In both side-bonded and μ-peroxo complexes the O-O bond length is generally between 1.4 and 1.5A while in superoxo complexes bond lengths vary from ~1.3 to 1.4A compared with the O-O bond length of 1.21 in the ground state dioxygen molecule.

Of the four different types of bonding shown in Figure 1, side-bonded peroxo complexes are the most common. Often they may be readily formed by merely bubbling oxygen through solutions of the lower valent group VIII metal complexes. Side-bonded peroxo complexes of the group V or VI metals are usually formed by reactions with hydroperoxides. μ-Superoxo complexes are the least common type and are exhibited only in the case of some Co(III) complexes. In fact, cobalt is the only metal for which all four bonding modes have been observed to date.

The d^7 cobalt(II) complexes react with molecular oxygen to give 1:1 end-bonded superoxo complexes which are stable in non-aqueous solvents at low temperatures provided the cobalt complex has a chelating ligand system of suitable structure, eq. 4 (7-11). More often the end-bonded superoxo complexes react with Co(II)

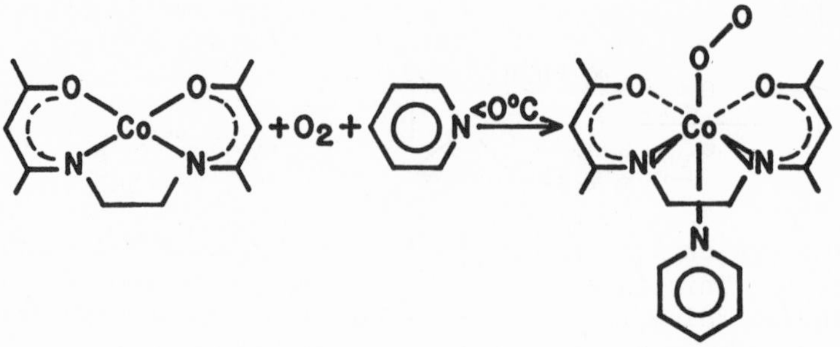

$$(4)$$

to form binuclear μ-peroxo complexes at varying rates (12-15). The μ-superoxo complexes are formed from μ-peroxo complexes in suitable oxidizing media (12). The sequence shown in eq. 5 shows this progression.

$$Co(II) \xrightarrow{O_2} Co(III)(O_2^-) \xrightarrow{Co(II)} Co(III)(O_2^{2-})$$

$$Co(III) \xrightarrow{-e^-} Co(III)(O_2^-)Co(III) \tag{5}$$

Cobalt(I) complexes, on the other hand, tend to form 1:1 peroxo complexes on reaction with dioxygen (16,17) eq. 6. They

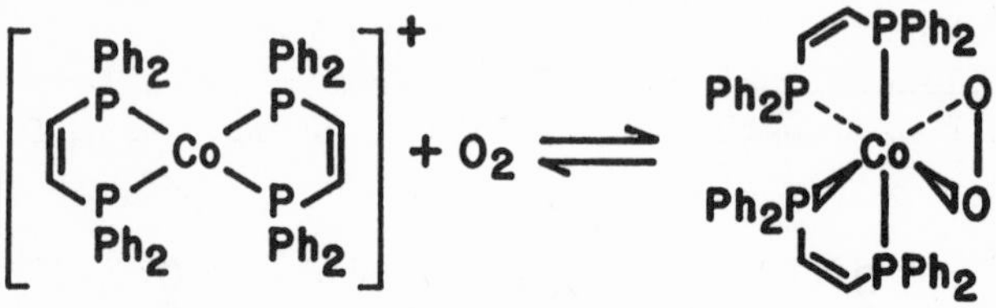

$$(6)$$

do so much faster than other group VIII d^8 complexes having comparable ligand systems. The reactivity of other group VIII d^8 complexes with molecular oxygen to give side-bonded dioxygen complexes is dependent on a number of factors (1-6). The metal center is important: reactivity increases in going from second to third row complexes (Os(0)>Ru(0), Ir(I)>Rh(I)). In general, reactivity is increased by strong σ-donor ligands and by highly polarizable ligands (3). Electron withdrawal from the metal whether by very strong πacceptor ligands, positive charge on the metal complex or other factors, lessens reactivity toward molecular oxygen (3). Group VIII d^8 complexes, Ni(0), Pd(0), Pt(0) also react readily with O$_2$ to give stable square planar dioxygen complexes (18-27), eq. 7. Neutral ligands are lost and in some cases are oxidized in the process.

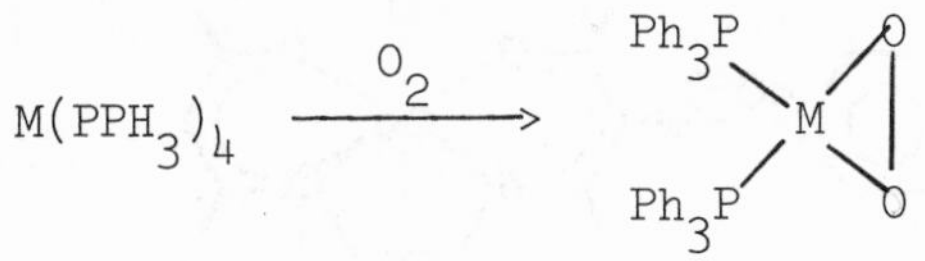

$$M(PPH_3)_4 \xrightarrow{O_2} \begin{array}{c} Ph_3P \\ \\ Ph_3P \end{array} M \begin{array}{c} O \\ | \\ O \end{array} \qquad (7)$$

In all cases oxygen appears to initially form a 1:1 complex with the metal. This interaction will result in an increase in oxidation and coordination numbers of the metal center by either one unit to give an end-bonded superoxo complex or by two units to give a side-bonded peroxo complex. Which oxidation reaction occurs depends on the number, arrangement and types of ligands, the d-electron configuration and the oxidation state and oxidation potential of the metal in the complex (1). It appears that the principle determining factors are the metal's oxidation state and the relative stabilities of its higher valences. Thus, low valent metal complexes tend to produce side-bonded peroxo species whereas metals in their normal oxidation states tend to give superoxo complexes (1).

Side-bonded peroxo complexes of groups V and VI are usually formed by reaction of the parent complex with hydrogen peroxide. Hydroperoxides are often present during catalytic oxidation of organic substrates and group V and VI metal side-bonded dioxygen complexes are capable of selectively oxidizing organic compounds. Examples of peroxo complexes known to react selectivity with organic substrates are complexes III and IV (28-30).

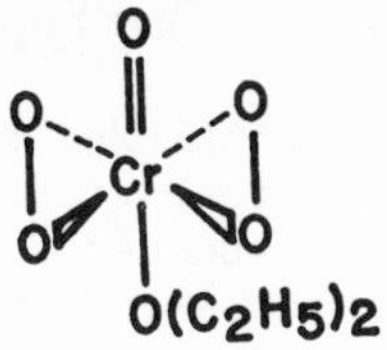

$$\begin{array}{c} O \\ \| \\ O \cdots Cr \cdots O \\ O \diagup \quad \diagdown O \\ O(C_2H_5)_2 \end{array} \qquad \begin{array}{c} O \\ \| \\ O \cdots Mo \cdots O \\ O \diagup \quad \diagdown O \\ HMPA \end{array}$$

III    IV

REACTIVITY OF COORDINATED OXYGEN

During the past decade a sizable number of reactions have been reported which involve the oxidation of substrates by dioxygen in the coordination sphere of the metal. These are

frequently reactions which occur sluggishly or not at all with
around state molecular oxygen. Consequently, dioxygen can be
considered to "activated" by coordination to a metal center.

End-bonded cobalt superoxo complexes have been used as models
for outer-sphere oxidation reactions. The ESR signal of the oxygen
complex of corrinoid [Co(II) (vitamin B_{12}] in the presence of small
amounts of one-electron reducing agents (hydroquinone, p-phenylene-
diamine, ascorbic acid, etc.) is replaced by the signal of the
radical of the oxidized substrate (7,31). In the presence of excess
reducing agent, Co(II) is regenerated and the reaction becomes
catalytic. Polarographic studies (32) show that paramagnetic cobalt
dioxygen complexes are more powerful one-electron oxidants than
free oxygen or the initial non-oxygenated cobalt(II) species.
Recently several well characterized cobalt dioxygen complexes have
been shown to exhibit oxidase and oxygenase activity (33). Both
ternary complexes of the type L-Co-O_2 (34) and Co-O-O-L (35,36),
have been proposed as intermediates.

$$\left[(en)_2Co \underset{O-O}{\overset{NH_2}{\diagup\diagdown}} Co(en)_2 \right]^{4+} \xrightarrow{SO_2} \left[(en)_2Co \underset{(SO_4)}{\overset{NH_2}{\diagup\diagdown}} Co(en)_2 \right]^{3+} \quad (8)$$

$$\left[(en)_2Co \underset{O-O}{\overset{NH_2}{\diagup\diagdown}} Co(en)_2 \right]^{4+} \xrightarrow{SeO_2} \left[(en)_2Co \underset{(SeO_4)}{\overset{NH_2}{\diagup\diagdown}} Co(en)_2 \right]^{3+} \quad (9)$$

Although instances of oxygen activation by μ-peroxo and
μ-superoxo complexes are few, the cobalt μ-superoxo complexes,
$[L_4Co(NH_2)(O_2)CoL_4]^{4+}$, react with SO_2 and SeO_2 to cleave the O-O
bond of coordinated dioxygen and form sulfato and selenato
linkages, eq. 8, 9 (37).

Group VIII metal side bonded dioxygen complexes have been
shown to be reactive toward a variety of organic and inorganic
substrates (37-42). Perhaps the most widely studied complex in
this regard is $[Pt(Ph_3P)_2(O_2)]$ (25-27, 43-47). Some of its
reactions are summarized in eq. 10 a-j. Group VI peroxo complexes also interact with reactive substrates. The molybdenum
peroxo complex $MoO_5(HMPA)$ has been shown to oxidize cyclohexene
to cyclohexene oxide, eq. 11 (30). It was postulated that a
cyclic peroxy intermediate was formed (30,48).

(10)

(11)

In addition to reactions of ligated dioxygen with organic substrates, reactions of coordinated substrates with dioxygen may have relevance to catalysis. In many cases, such as reactions of SO_2 complexes with dioxygen, the results are the same as those reviewed above suggesting the possibility of common intermediates (25, 38, 49-51). Reactions of coordinated CO also give carbonato complexes except in the case of rhodium complexes which tend to form coordinated or free CO_2 (52,53).

Since metal hydrides and metal alkyls are often intermediates of catalytic reactions of unsaturated hydrocarbons, reactions of these species with O_2 are of interest. Cationic hydrido complexes of Ir, Rh, Ru and Os react with molecular oxygen to insert oxygen between the metal and the hydrido ligand, eq. 12 (54, 55). Oxygen has also been shown to insert oxygen between the cobalt-alkyl bond in cobaloximes, eq. 13 (56). Iridium alkylperoxides have been reported (57), but they were formed by reaction of

$$[Rh(NH_3)_5H]^{2+} \ + \ O_2 \ \longrightarrow \ [Rh(NH_3)_4(OH)(OOH)]^{-1} \qquad (12)$$

$$\text{Oxime Co(II)}\text{---}R \ + \ O_2 \ \xrightarrow{\ h\nu\ } \ \text{Oxime Co(II)} \text{---} OOR \qquad (13)$$

iridium(I) complexes with alkyl hydroperoxides, eq. 14. Both CO_2 and Ph_3PO were also formed in this reaction.

$$IrL_2(CO)_X \ + \ ROOH \ \longrightarrow \ IrL_2(CO)X(OOR)_2 \qquad (14)$$

$$L=PPh_3, \ AsPh_3, \ PPh_2Me; \ X=Cl, \ Br; \ R\text{-}\underline{t}\text{-}Bu, \ PhMe_2$$

$[Rh(CO)Cl(PPh_3)_2]$ in benzene under nitrogen consumed one equivalent of Bu^tOOH to give CO_2 (0.95 equiv.) (58) eq. 15. No Ph_3PO was detected, and it is of interest that the carbonyl was

$$Rh(PPh_3)_2(CO)Cl \ + \ ROOH \ \longrightarrow CO_2 \ + \ [Rh(PPh_3)_2Cl]_2 \qquad (15)$$

oxidized in preference to a phosphine ligand. A cobalt(III) tetraarsine complex, on the other hand, reacts with hydrogen peroxide to give a side-bonded peroxo complex (17), eq. 16.

$$[Co(III)(tetars)] \xrightarrow{H_2O_2} \left[\begin{array}{c} As \\ As-\!|\!-As \\ Co \\ As\diagup\ \ |\ \diagdown O \\ O \end{array} \right]^+ \qquad (16)$$

$$[Co(tetars)O_2]^+$$

CATALYTIC OXIDATION REACTIONS

Since it has been established that molecular oxygen can be readily brought into the coordination sphere of a metal complex and that ligated dioxygen is a more reactive and selective reagent than ground state molecular oxygen, it is reasonable to suppose that coordination catalysis might play a major role of considerable synthetic importance. Provided that the oxidized substrate is a more weakly held ligand than the original substrate, catalysis should be possible. Thus, oxidation of phosphines to phosphine oxides, sulfides to sulfoxides, isocyanides to isocyanates and olefins to epoxides might all be possible. With the advent of homogeneous catalysis as a powerful synthetic tool in the 1960's, it was hoped that these and even more interesting applications might be uncovered.

It has been shown, however, that homogeneous catalysis of oxidation is seldom straightforward, and complex patterns of reactivity are the rule in most instances. One reason is that reaction pathways leading to the generation of organic free radicals are often possible and organic peroxides may form. The metal catalyzed decomposition of organic hydroperoxides to give organic radicals and the subsequent radical initiated autoxidation of the substrate often compete with coordination catalysis. Metal complexes vary widely in their ability to decompose hydroperoxides, and to some extent in the manner in which they decompose them. Selective catalytic conversion of an organic hydroperoxide to stable oxidation products in the presence of certain metal complexes affords convenient routes to useful chemicals. Unselective decomposition to radicals which initiate further unselective reactions often occurs, however.

In the remainder of this review, we will attempt to assess the progress of the past decade in achieving selective catalytic transfer of molecular oxygen to organic substrates in a catalytic

manner. We will limit discussion to those substrates which should
be particularly well suited to coordination catalysis, i.e., those
having systems of p or π-electrons by which they might react
readily with the metal complexes used as catalysts.

CATALYTIC OXIDATION OF PHOSPHINES TO PHOSPHINE OXIDES

Many transition metal complexes have been found to catalyze
the oxidation of organic phosphines to phosphine oxides. From
the bulk of the evidence it appears that some form of catalytic
oxygen activation occurs. Wilke, Schott and Heimbach (24) showed
that complexes of the type: [M(PPh$_3$)$_4$] catalyze the oxidation of
triphenylphosphine to triphosphine oxide in toluene solution.
This reaction has been extensively studied using [Pt(PPh$_3$)$_4$] as
the catalyst by Halpern and co-workers who have elucidated many
aspects of the reaction mechanism (26,27,59,60). These authors
have shown that the predominant species in benzene solution is
[Pt(PPh$_3$)$_3$] which adds dioxygen to give [Pt(PPh$_3$)$_2$(O$_2$)]. The
platinum dioxygen complex reacts with external triphenylphosphine
to form Ph$_3$PO and regenerate the catalytically active species,
[Pt(PPh$_3$)$_3$]. The kinetics of the individual steps as well as of
the overall catalytic reaction have been examined and are in
agreement with the mechanism shown in eq. 17. More recent
studies (61) indicate that oxygen atom transfers which were

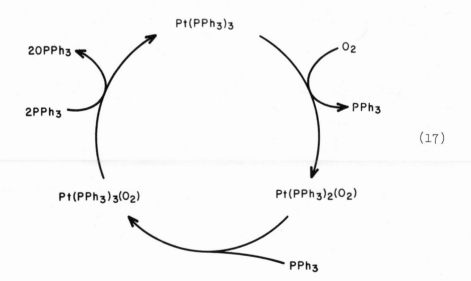

$$(17)$$

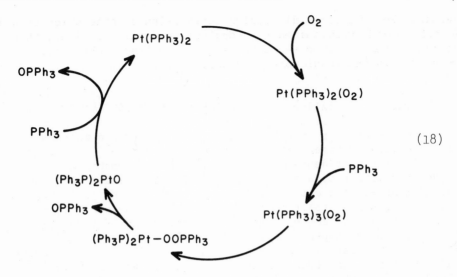

(18)

originally postulated to occur within the coordination sphere probably are not operative. Stern (62) has suggested another mechanistic scheme, eq. 18, to describe this reaction.

The catalytic oxidation of triphenylphosphine to triphenylphosphine oxide has also been investigated in detail using ruthenium complexes as catalysts (63-67). Among the catalytically active ruthenium complexes which have been investigated is [Ru(NCS)(CO)(NO)(PPh$_3$)$_2$]. The mechanism shown in eq. 19 and 20, is consistent with kinetic data. In this instance oxygen atom transfer within the coordination sphere was proposed as the slow step. Using RuCl$_2$(PPh$_3$)$_3$, James and co-workers (67) have

$$O_2 + Ru(NCS)(CO)(NO)(PPh_3)_2 \underset{}{\overset{K_1}{\rightleftharpoons}} Ru(NCS)(NO)(PPh_3)_2(O_2) + CO \quad (19)$$

$$Ru(NCS)(NO)(PPh_3)_2(O_2) \underset{}{\overset{K_2 PPh_3}{\rightleftharpoons}} Ru(NCS)(NO)(PPh_3)_3(O_2)$$

$$\uparrow O_2, \text{ fast} \qquad\qquad \downarrow \text{slow} \qquad (20)$$

$$Ru(NCS)(NO)(PPh_3)_2 \xleftarrow[- 2OPPh_3]{PPh_3} Ru(NCS)(NO)(PPh_3)(OPPh_3)_2$$

recently shown that at low temperatures triphenylphosphine oxide is formed by external attack of PPh_3 on $[RuCl_2(PPh_3)_2(O_2)]$ whereas at ambient temperatures O-atom transfer to phosphine may occur within the complex $[RuCl_2(PPh_3)_2(O_2)]$.

Rhodium(I) and Iridium(I) complexes have been used in a similar manner to oxidize triphenylphosphine (68-73). Mechanisms involving oxygen-atom transfer within the coordination sphere have been proposed in these cases as well. Cobalt complexes have also been used as catalysts in phosphine oxidation (74-76). Schmidt and Yoke (74) have found that the sole produce of the selective quantitative oxidation of $[CoCl_2(PEt_3)_2]$ is the phosphine oxide complex, eq. 21.

$$CoCl_2(PEt_3)_2 + O_2 \longrightarrow CoCl_2(OPEt_3)_2 \qquad (21)$$

In its initial stages the reaction is first order in oxygen and first order in cobalt complex, eq. 22. Free radical initiators or inhibitors have no effect on the rate of oxygen consumption.

$$d[O_2]/dt = k[CoCl_2(PEt_3)_2]Po_2 \qquad (22)$$

In any case a dissociative mechanism is precluded since radical initiated autoxidation of uncoordinated phosphine gives mixed $R_nPO(OR)_{3-n}$ products unselectively. The mechanism suggested involves the formation and rearrangement of a cobalt dioxygen complex, eq. 23, 24 possibly via dissociative oxygen insertion. During the reaction, redistribution to give a mixed ligand complex, eq. 25, was observed with an equilibrium constant of about 10. Although this system appears to possess all of the attributes necessary for the selective catalytic oxidation of trialkylphosphines to phosphine oxides, the catalytic activity of these cobalt complexes was not reported.

$$CoCl_2(PEt_3)_2 + O_2 \rightleftharpoons CoCl_2(PEt_3)_2(O_2) \qquad (23)$$

$$CoCl_2(PEt_3)_2(O_2) \longrightarrow CoCl_2(OPEt_3)_2 \qquad (24)$$

$$CoCl_2(PEt_3)_2 + CoCl_2(OPEt_3)_2 \rightleftharpoons 2CoCl_2(PEt_3)(OPEt_3) \qquad (25)$$

A novel and selective type of oxygen activation which has not been mentioned yet has been reported by Barral, Bocard, Seree de Roch and Sajus (77-79). These authors have shown that the molybdenum oxo complexes: $[MoO_2(S_2CNR_2)_2]$ (R=ethyl, n-propyl or isobutyl) (80) catalyze the selective oxidation of tri-n-butylphosphine to the corresponding phosphine oxide. Again, the fact that $OPBu_3$ is selectively formed from PBu_3 without products such as $OPBu_2(OBu)$, argues against a dissociative radical auto-xidation mechanism and implies coordination catalysis. The kinetics of the probable steps of the catalytic reaction have been studied individually and together with kinetics of the overall process, suggest the oxidation scheme described in eq. 26.

$$
\begin{array}{c}
PR'_3 \quad\xrightarrow{k_r}\quad OPR'_3 \\
\\
MoO_2(S_2CNR_2)_2 \quad + \quad MoO(S_2CNR_2)_2 \quad\xrightleftharpoons{K}\quad Mo_2O_3S_2(CNR_2)_4 \\
\\
+\ 1/2\ O_2\ k_o
\end{array}
\qquad (26)
$$

We have considered the oxidation of organic phosphines in some detail because as a group of reactions, they constitute the most widely studied examples of coordination catalysis of oxidation. It seems likely that in the presence of many different metal complexes, oxygen is indeed catalytically activated and that selective oxidations of this type might serve as a model for other homogeneous catalytic oxidations.

OXIDATIONS OF ORGANIC ARSENIC, SULFUR & NITROGEN COMPOUNDS

Many of the same complexes which catalyze phosphine oxidation are also effective for oxidation of arsines to arsine oxides (64, 66, 81). The ruthenium complex, $[Ru(NCS)(NO)(PPh_3)_2O_2]$, for example has been found to be an efficient catalyst for the oxidation of triphenylarsine (64); the initial rate of oxidation being three or four times greater than the corresponding rate for triphenylphosphine oxidation under the same conditions.

Examples of the direct O_2 oxidation of dialkyl sulfides catalyzed by metal complexes are very few in number. Sulfides are oxidized to sulfoxides in the presence of $[VO(acac)_2]$ (82), and it has recently been reported (83) that dibenzothiophene may be oxidized to dibenzothiophene-5,5-dioxide in the presence

of $[Ru_3(CO)_{12}]$, $[Ru(acac)_3]$ and $RuCl_3$ at elevated temperatures.
Henbest and Trocha-Grimshaw have shown that sulfoxides are
oxidized to sulfones in the presence of catalytic quantities of
rhodium and iridium complexes (84, 85).

Although little work has appeared concerning the reaction
of molecular oxygen with dialkyl sulfides catalyzed by metal
complexes, the homogeneous catalytic oxidation of dialkyl sulfides
with hydroperoxides is accomplished with ease in the presence
of zirconium, vanadium, molybdenum or titanium complexes (86-90).
The selectivity of this reaction is demonstrated by reaction 27
in which alkyl n-butyl sulfide is oxidized to the sulfone without
appreciable oxidation of the olefinic group (89, 90). Dialkyl
sulfide can also be oxidized selectively to the sulfone without
oxidation of the unsaturated side chain (89). Coordination of
the sulfur is apparently strong enough to exclude the olefinic

$$
\begin{array}{c}
CH_3CH_2CH_2CH_2 \\
\diagdown \\
S \xrightarrow[MoO_2(acac)_2]{t-BuOOH} \\
\diagup \\
CH_2=CHCH_2
\end{array}
\qquad
\begin{array}{c}
CH_3CH_2CH_2CH_2 \\
\diagdown \\
SO_2 \\
\diagup \\
CH_2=CHCH_2
\end{array}
\qquad (27)
$$

ligand from the reactive center since in the absence of the
sulfur, olefins are readily epoxidized by t-BuOOH in the presence
of molybdenum complexes. The product of hydroperoxide oxidation
of sulfides depends upon reaction conditions (88). Sulfoxides
are obtained at temperatures below 50°C using excess sulfide
whereas sulfones are the predominant product using excess
hydroperoxide. Little work has been reported concerning metal
catalyzed oxygenation of secondary and tertiary amines with
dioxygen. Primary amines couple to give azo compounds in the
presence of copper complexes (94-101). The mechanism of this
reaction is not clearly understood. The oxidation of alkyliso-
cyanides to isocyanates is catalyzed by a variety of metal complexes (21,60,93) including $[Ni(t-BuNC)_4]$, $[Ni(t-BuNC)_2(O_2)]$,
$[RhCl(Ph_3P)_3]$, $[Co(1,5-C_8H_{12})_2]$, and $[Ni(1,5-C_8H_{12})_2]$, eq. 28.

$$
Ni(1,5-C_8H_{12})_2 \xrightarrow{t-BuNC} Ni(t-BuNC)_4 \xrightarrow{O_2}
$$

$$
\begin{array}{c}
t-BuNC \diagdown \quad \diagup O \\
Ni \\
t-BuNC \diagup \quad \diagdown O
\end{array}
\xrightarrow{t-BuNC} t-BuNCO + Ni(t-BuNC)_4
\qquad (28)
$$

A variety of organic nitrogen compounds undergo metal catalyzed oxidative cleavage reactions with ease. Copper(I) complexes are particularly effective catalysts for cleavage reactions. Aldehyde enamines for example are selectively cleaved using catalytic quantities of cuprous chloride under extremely mild conditions (102,103). The mild conditions under which the catalytic reaction occurs permits the rapid and selective cleavage of the enamine double bond in the presence of other sites of unsaturation in an organic molecule. A particularly elegant example of the selectivity of this reaction is the quantitative oxygenation of V to give progesterone, VI, and N-formylmorpholine, VII, eq. 29.

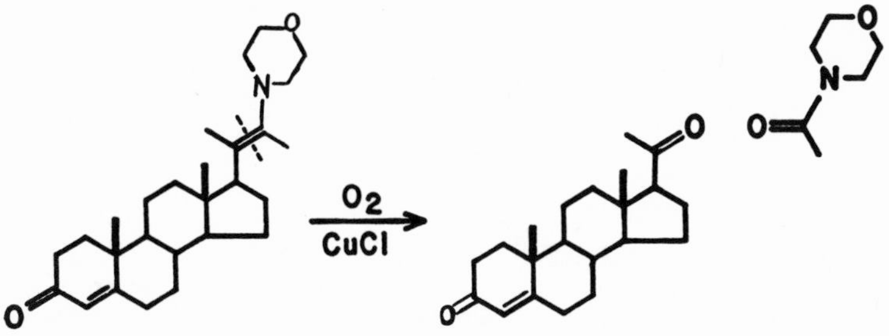

$$ (29) $$

A kinetic study revealed the oxygenation rate to be first order in V after an induction period, but also dependent on the cuprous chloride concentration. The radical inhibitor, 2,6-di-t-butyl-p-cresol did not retard the reaction but strongly coordinating ligands which can compete with the enamine inhibited reaction completely. Recent studies (104-106) have extended the utility of the enamine oxygenation reaction and have examined the effects of substituents on this reaction.

The catalyst formed when oxygen is added to copper(I) chloride in pyridine has found considerable utility. The preparation of unsaturated nitriles from unsaturated amines can be accomplished in a catalytic manner using this system (107,108). The oxidative cleavage of o-phenylenediamine takes place at room temperature, eq. 30 (108). The catalytic oxidation of dihydrazones to acetylenes (109) also proceeds smoothly the presence of CuCl in pyridine, eq. 31. Although the mechanisms of these reactions are still obscure, evidence has recently been presented (110) for the formation of copper peroxo complexes of the formula $(py)_n CuOOCu(py)_n$ from oxygenation of pyridine solutions

$$\text{(aniline)} + O_2 \xrightarrow[\text{pyridine}]{\text{CuCl}} \text{(dicyanobutadiene)} \quad (30)$$

$$\begin{array}{c} R \quad R \\ \diagdown \; \diagup \\ C \quad C \\ \| \quad \| \\ N \quad N \\ \diagup \quad \diagdown \\ NH_2 \quad NH_2 \end{array} + O_2 \xrightarrow[\text{pyridine}]{\text{CuCl}} RC \equiv CR + 2N_2 + 2H_2O \quad (31)$$

of copper(I) chloride. That the peroxo oxygen is activated was
shown by the facile oxidation of SO_2 by this complex to give
copper sulfate (110).

Cobalt complexes are also capable of catalyzing oxidative
cleavages of some nitrogen compounds. Cobalt(II) was found to
be uniquely active among a series of metal tetrasulfophthalo-
cyanine complexes for the oxidative cleavage of hydrazine (111).
This catalytic activity was ascribed to the ability of the Co(II)
complex to reversibly bind molecular oxygen. Kinetic studies
suggest the mechanism shown in eq. 32.

$$\text{CoTSP} \underset{\text{CoTSP}(N_2H_4)}{\overset{\text{CoTSP}(O_2)}{\rightleftharpoons}} \text{CoTSP}(N_2H_4)(O_2) \longrightarrow \text{CoTSP} + \quad (32)$$
$$N_2 + H_2O$$

Cobalt(III) acetate has been shown to possess activity in the
oxidative cleavage of diamines such as benzidene and o,o'-di-
anisidene to the corresponding quinones in acetic acid. A
radical mechanism is probably involved in this case (112).

Salicylaldehydethylenediminato cobalt(II), [Co(salen)], and
[CoTPP] also catalyze the oxidative cleavage of substituted
indoles, (112a,b), eq. 33.

An interesting stoichiometric oxidative cleavage reaction
of an oxime with a palladium dioxygen complex has been observed
(113). The palladium dioxygen complex, $[Pd(PPh_3)_2(O_2)]$, has

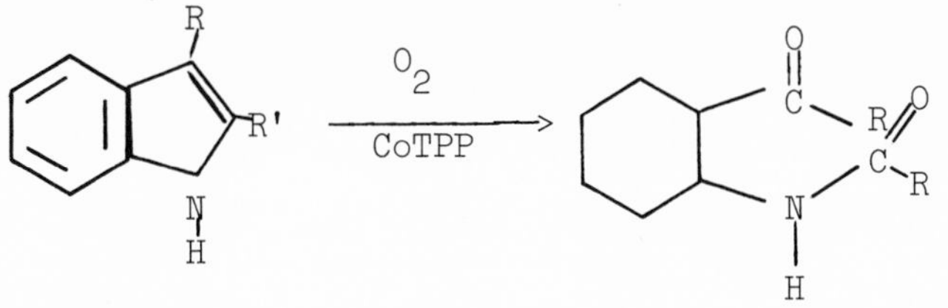

$$(33)$$

been shown to rapidly deoximate a variety of ketoximes in benzene at 25°C to give nearly quantitative yields of ketones. A 1,3-dipolar cycloaddition of the dioxygen complex to the ketoxime was proposed, eq. 34.

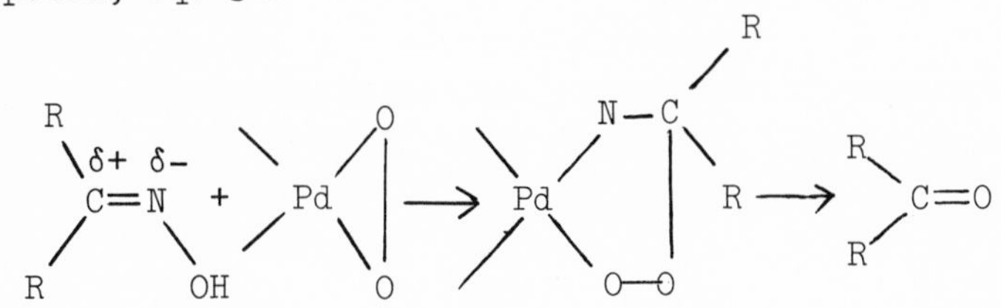

$$(34)$$

Although tertiary amines are not smoothly oxidized by molecular oxygen using metal complexes as catalysts, they are oxidized selectively to amine oxides by organic hydroperoxides in the presence of group VB and VIB transition metal complexes (114-119). The kinetics of the oxidation of a series of 3-substituted pyridenes showed that reactions were first order in both hydroperoxide and amine and a linear relationship was found between reactivities and Taft σ constants (118). Reaction rate was found to increase with decrease in the electron density on the reaction center; $\rho = +1.2$.

Equation 35 was proposed as a schematic representation of the reaction mechanism.

Hydroperoxides have been used to carry out a variety of other oxidations of nitrogen compounds in the presence of group IVB to VIB metal complexes. Primary aromatic amines are oxidized to Azoxy compounds in the presence of titanium complexes (115,119), eq. 36 or to nitro compounds when vanadium and molybdenum complexes are used (120). In contrast to the oxidation of 3° amines, the effect of substituents upon reaction rate in this case corresponds to a reaction involving an electron deficient transition state in that electron withdrawing groups decrease the rate. Straight line correlations between the log of the relative rate and Hammett σ or Brown σ+ constants were obtained with ρ values of

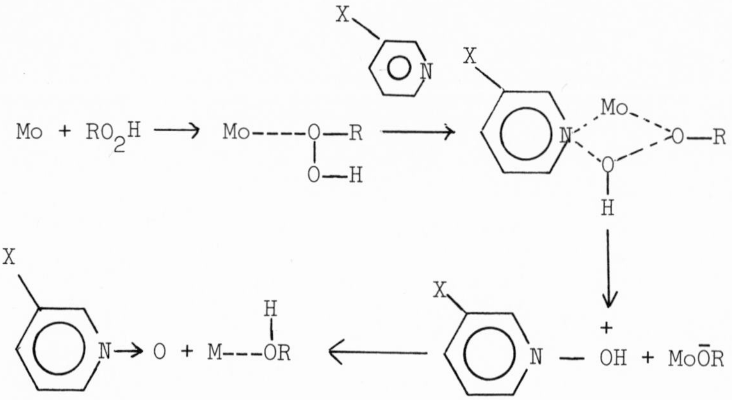

(35)

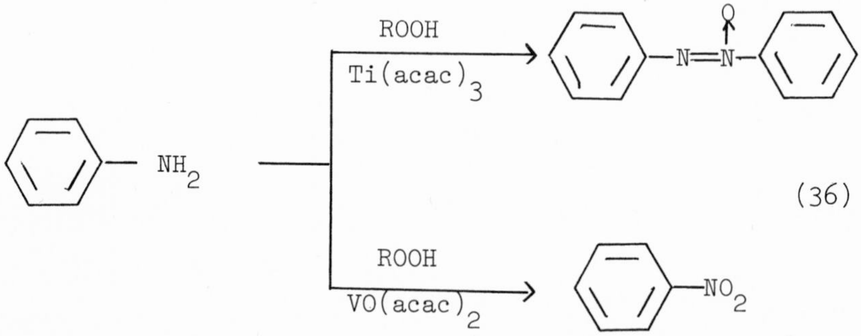

(36)

-1.42 and -1.97, respectively. The rate determining step in
this reaction was postulated to involve attack by nitrogen on
coordinated hydroperoxide, eq. 37 (120).

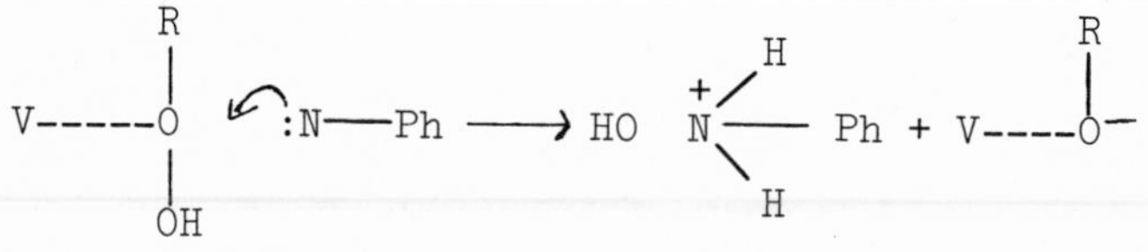

(37)

Other catalytic hydroperoxide oxidations of nitrogen compounds which have been reported include the oxidation of primary
amines to oximes (121,122), eq. 38, imines to oxaziridines (117),
eq. 39, nitrosobenzene to nitrobenzene (123), eq. 40, nitrosoamines
to nitroamines (117), eq. 41, and azobenzene to azoxybenzenes
(124), eq. 42.

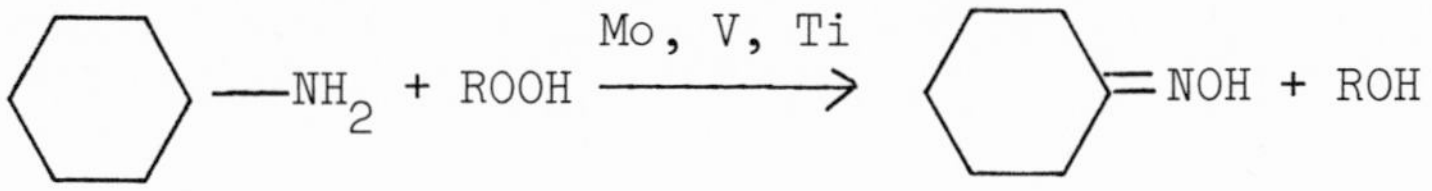

$$\text{C}_6\text{H}_{11}\text{—NH}_2 + ROOH \xrightarrow{\text{Mo, V, Ti}} \text{C}_6\text{H}_{10}{=}\text{NOH} + ROH \qquad (38)$$

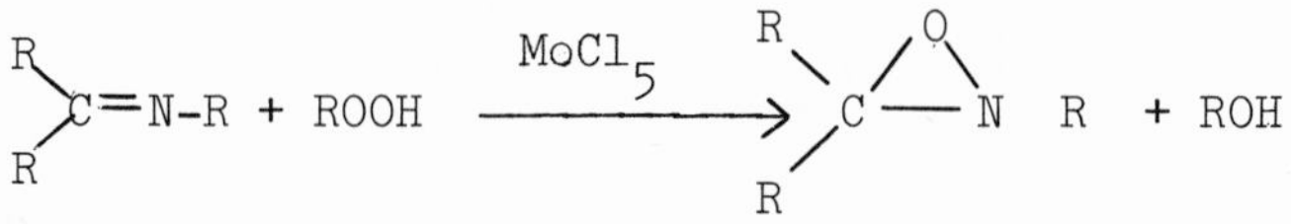

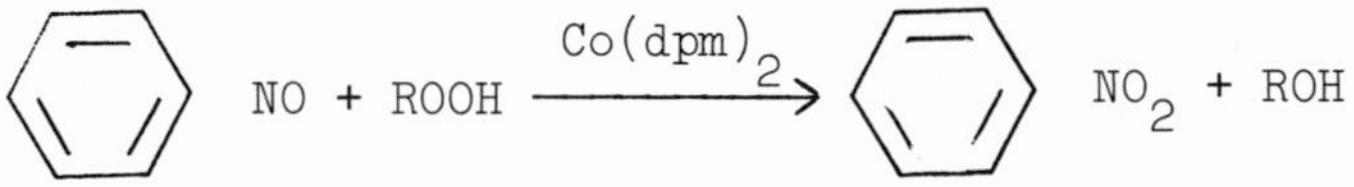

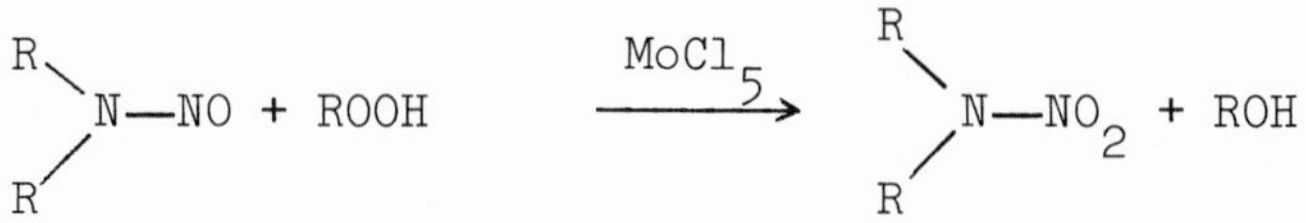

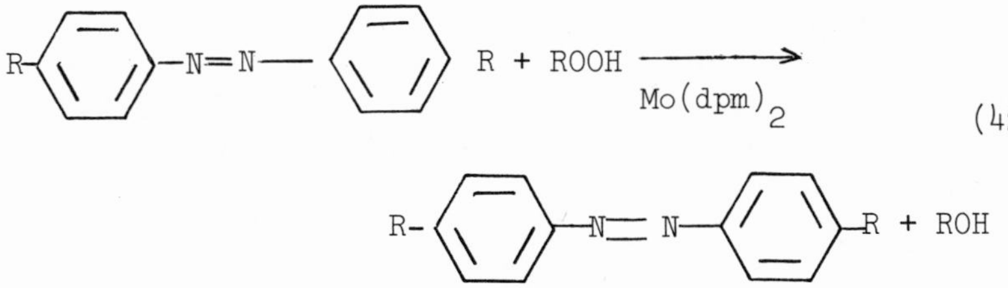

$$\underset{R}{\overset{R}{>}}\text{C}{=}\text{N-R} + ROOH \xrightarrow{\text{MoCl}_5} \underset{R}{\overset{R}{>}}\text{C}{-}\text{N} \; R + ROH \qquad (39)$$

$$\text{C}_5\text{H}_5\text{N} \cdot NO + ROOH \xrightarrow{\text{Co(dpm)}_2} \text{C}_5\text{H}_5\text{N} \cdot NO_2 + ROH \qquad (40)$$

$$\underset{R}{\overset{R}{>}}\text{N—NO} + ROOH \xrightarrow{\text{MoCl}_5} \underset{R}{\overset{R}{>}}\text{N—NO}_2 + ROH \qquad (41)$$

$$R{-}\text{Ar}{-}N{=}N{-}\text{Ar} \; R + ROOH \xrightarrow{\text{Mo(dpm)}_2} \qquad (42)$$

$$R{-}\text{Ar}{-}N{=}N{-}\text{Ar}{-}R + ROH$$

OXIDATIONS OF CO, ALDEHYDES, KETONES & ALCOHOLS

Carbon monoxide is readily oxidized in the coordination sphere of a number of transition metal complexes. In many cases the product of reaction is a carbonate complex which is formed irreversibly, thus precluding the possibility of a catalytic transformation. Some rhodium carbonyl complexes, on the other hand, react with dioxygen in aromatic solvents to give a comples containing coordinated CO_2 (53), eq. 43. Since rhodium carbon dioxide complexes

$$\text{Rh}_2(\text{PPh}_3)_4(\text{CO})_4 + O_2 \longrightarrow \text{Rh}_2(\text{PPh}_3)_3(\text{CO})_2 + \text{OPPh}_3 \qquad (43)$$
$$+ \; CO$$

are not as stable as the corresponding carbonato complexes with respect to CO, catalysis is possible. Kiji and Furukawa reported

that both [RhCl(CO)(PPh$_3$)$_2$] and [RhCl(CO)(Me$_2$SO)$_2$] catalyzed the
oxidation of CO in benzene, ethanol, or dimethylsulfoxide (125).
The rhodium cluster compounds, [Rh$_6$(CO)$_{16}$], is an even more
efficient catalyst for this reaction (126). In contrast to
[Pt(PPh$_3$)$_2$(O$_2$)] which forms a carbonate complex with CO, the nickel
complex, Ni(t-BuNC)$_2$(O$_2$) , reacts with CO at 20°C to give
[Ni(CO)$_2$(t-BuNC)$_2$] and CO$_2$ (49). Catalytic activity, however was
not reported.

A variety of metal complexes are known to catalyze oxidation
of CO to CO$_2$ in aqueous media (127-151). The mechanism of CO$_2$
formation in these cases, however, may not involve oxygen atom
transfer to coordinated CO. One mechanism which may be operative
in some cases involves attack by ater on CO within the coordination
sphere to give CO$_2$ and a reduced metal complex. The function of
oxygen may, therefore, be mrely to oxidize the metal complex back
to the catalytically active state. In other instances oxygen
activation and transfer to coordinated CO have been postulated in
aqueous media. Recently Likholobov, et al., have shown (195 that
palladium phosphine complexes catalyze the oxidation of CO at 40 C
in aqueous dioxane, provided that an acid, HX (X=CH$_3$CO$_2$, CF$_3$CO$_2$,
NO$_3$, BF$_4$, ClO$_4$, etc). whose anion is a weakly coordinating ligand,
is present. The authors suggest a mechanism involving; a)
coordination of dioxygen; b) oxygen atom transfer to CO; c)
decomposition of the carbonate complex by HX; and d) reduction of
Pd(II) to Pd(0) by CO, eq. 45.

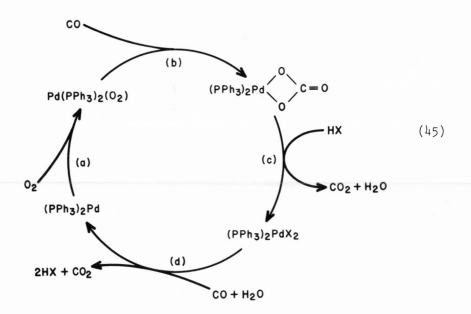

$$(45)$$

In the presence of dehydrating agents such as trialkyl ortho-
formates, the $PdCl_2/CuCl_2$ redox system catalyzes the reaction of
CO, O_2 and ROH to form dialkyl oxalates (152, 153). The reaction
pathway proposed for oxalate formation is shown in eq. 46.
Copper chloride and oxygen reoxidize the reduced palladium to
Pd(II) and water and the dehydrating agent removes the water
which is produced.

$$\begin{array}{c} ROH \\ \diagdown\diagup \\ Pd \\ \diagup\diagdown \\ OC \end{array} \xrightarrow{CO} \begin{array}{c} ROC{\nwarrow}^{O}\;H \\ \diagdown\;|\diagup \\ Pd \\ \diagup\diagdown \\ OC \end{array} \xrightarrow{ROH} \begin{array}{c} RO-C{\nearrow}^{O} \\ | \\ RO-C{\searrow}_{O} \end{array} + PdH_2 \qquad (46)$$

Aldehydes readily undergo oxidation to the corresponding acid
on standing in air (154). Reaction generally proceeds by a chain
mechanism, eq. 47-49, to give peracid which often reacts further
to give the corresponding carboxylic acid, eq. 50. The effect of
metal ions on aldehyde oxidation is chiefly to assist in the
formation of radical species which initiate reaction, eq. 51.

$$PhCHO \; + \; R\cdot \; \longrightarrow \; Ph\dot{C}O \; + \; RH \qquad (47)$$

$$Ph\dot{C}O \; + \; O_2 \; \longrightarrow \; PhCOO_2\dot{} \qquad (48)$$

$$PhCOO_2\dot{} \; + \; PhCHO \; \longrightarrow \; PhCOO_2H \; + \; Ph\dot{C}O \qquad (49)$$

$$PhCOO_2H \; + \; PhCHO \; \longrightarrow \; 2PhCO_2H \qquad (50)$$

$$M^{n+} \; + \; RCHO \; \longrightarrow \; R\dot{C}O \; + \; M^{n+1} \; + \; H^+ \qquad (51)$$

Mixtures of peracids and carboxylic acids are usually produced in metal ion promoted autoxidation of aldehydes (155-161),
the selectivity of this reaction is dependent on the metal center
and the ligand system. The selectivity to peracetic acid from
acetaldehyde in the presence of naphthenate complexes decreased
in the order: Fe, Co, none>>Ni, Cr, V, Cu>>Mn. Thus, manganese
complexes are particularly effective if high selectivities to the
acid are desired (162) and iron or cobalt should be used for high
selectivity to peracids. Both the activity and selectivity of
copper(II) catalyzed oxidation of aldehydes to peracids is
enhanced by addition of 2,2' bipyridyl. Recent work (163) has
shown that cobalt(II) tetraphenylprophyrins catalyze the autoxidation of acetaldehyde giving peracid quantitatively. Para-sub-
stituents on the phenyl groups of the porphyrinic ligand influenced
catalytic activity and the effect of the added base was large.
From the experimentally derived rate equation and ESR studies, it

was concluded that an oxygen molecule activated by electron transfer from cobalt(II) was able to abstract the hydrogen of
acetaldehyde to initiate autoxidation as shown in eq. 52.

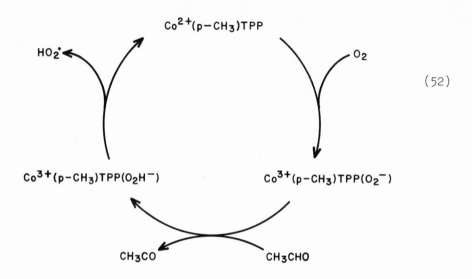

(52)

In a recent report (164) it has been shown that the dioxygen
complexes formed from several square planar d^8 or d^{10} metal complexes catalyzed the oxidation of benzaldehyde to a mixture of
perbenzoic and benzoic acids. Reaction was inhibited by free
phosphine and markedly retarded by perbenzoic acid while benzoic
acid had no effect. The authors (164) noted that it is unlikely
that peracids should have a retardation effect if this reaction
were a radical chain process since peracids are known to act as
initiators or reoxidizing agents. They propose a mechanism,
eq. 53-56, which is in agreement with the experimentally determined
rate equation.

$$Pt(PPh_3)_2(O_2) \; + \; PhCHO \; \rightleftharpoons \; Pt(PPh_3)_2(O_2)(PhCHO) \quad (53)$$

$$Pt(PPh_3)_2(O_2)(PhCHO) \; \rightleftharpoons \quad (54)$$

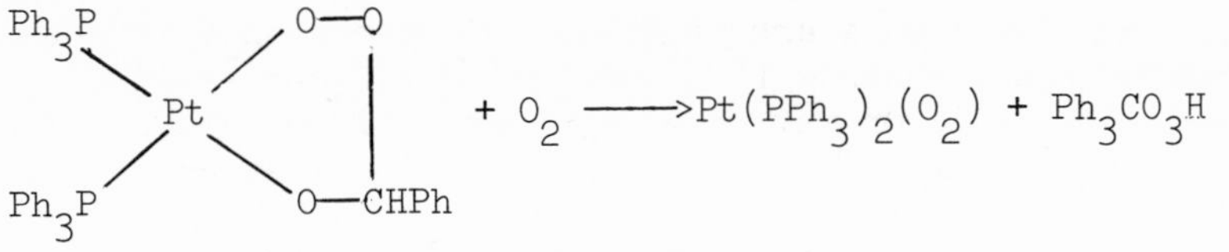

$$+ O_2 \longrightarrow Pt(PPh_3)_2(O_2) + Ph_3CO_3H \quad (55)$$

$$PhCO_3H + PhCHO \rightleftharpoons Ph\overset{O}{\underset{\|}{C}} - O - O - \overset{OH}{\underset{|}{C}H} -Ph \longrightarrow 2PhCOOH \quad (56)$$

The oxidation of ketones in the presence of metal complexes
to give carboxylic acids and aldehydes is well known (126, 154,
165-192). Komissarov and Denisov (265-267) have shown that an
iron(III)-o-phenanthroline complex and a copper(II) pyridine complex catalyze this reaction. The mechanism shown in eq. 57. was
proposed. Kinetic H/D exchange reactions support the presence of
a rate determining enolization step. Although ketone oxidation is
most probably radical initiated, the metal center still exerts
control over the predominant reaction pathway as shown in eq. 58
(183-186).

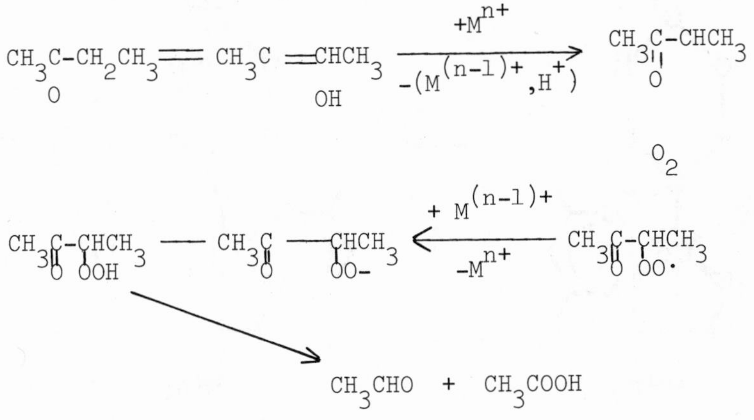

$$(57)$$

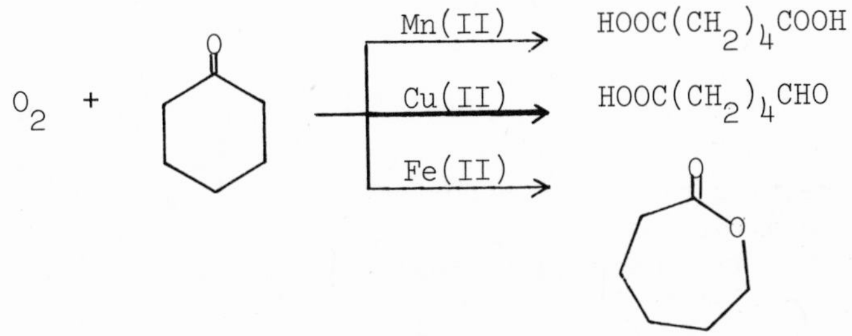

$$(58)$$

Seconadary alcohols are oxidized in the presence of metal complexes to yield ketones (193) and primary alcohols give aldehydes (194, 195) or may react further to form carboxylic acids (196, 197). Radicals are probably involved in most cases, however, evidence has been presented (198) that the oxidation of phenyl-methylcarbinol catalyzed by CoBr$_2$ takes place without formation of free radicals. It was found that while 2-propanol solutions of [Co(salen)] reacted reversibly with dioxygen, in the presence of triphenylphosphine irreversible uptake of oxygen occurred (199). Kinetic investigations suggested the formation of a labile pre-reaction complex of [Co(salen)] with triphenylphosphine, oxygen and the alcohol. The rate determining step was postulated to involve an outer sphere hydrogen migration from the alcohol to the dioxygen ligand (199), eq. 59.

(59)

In an interesting series of reports (200-202) ferrous ion-hydrogen peroxide oxidation of cyclohexanol in acetonitrile has been shown to occur with remarkable regioselectivity to give predominantly cis-1,3-cyclohexanediol. While these results appear inconsistent with a mechanism involving free hydroxyl radical, they are in accord with a scheme mediated instead by an iron species (202), eq. 60.

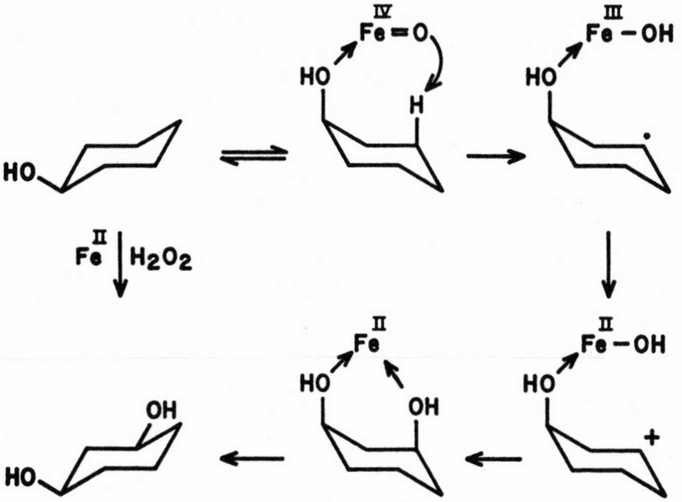

$$(60)$$

OXIDATION OF OLEFINS

Of the many substrates which have been oxidized in the presence of transition metal complexes, one of the most extensively studied groups of compounds has been the olefinic hydrocarbons. The obvious incentive for the pursuit of this research is the identification of catalysts which could convert the abundant olefinic hydrocarbons into more valuable oxygen-containing derivatives under mild conditions in high selectivity. Since hydrocarbon-soluble complexes of the transition metals have been successfully applied to the catalytic addition of other small molecules to olefinic substrates, attempts have been made to activate and catalytically transfer dioxygen to olefins in a similar manner. It has been difficult, however, to exclude other pathways by which oxygen can interact with olefins and to achieve selective reactions. Hydroperoxide intermediates are formed wherever possible and free radical pathways usually result. In cases where selective oxidations are achieved it is often hard to decide whether products arise via coordination catalysis or autoxidation pathways.

a) Hydroperoxide Reactions

Because of the prevalence of hydroperoxides in hydrocarbon oxidation a brief consideration of their reactivity should be considered before discussing olefin oxidation. Tertiary hydroperoxides are readily decomposed in the presence of catalytic

quantities of metal complexes (203), eq. 61. Regardless of the
catalyst used the products are the alcohol and molecular oxygen.
It has been well established that, products arise _via_ the general
route shown in eq. 62-64 (349-353).

$$2R_3COOH \xrightarrow{\text{M}} 2R_3COH + O_2 \tag{61}$$

$$M^n + ROOH \longrightarrow M^{n+1} + RO\cdot + OH^- \tag{62}$$

$$M^{n+1} + ROOH \longrightarrow M^n + ROO\cdot + H^+ \tag{63}$$

$$2ROO\cdot \longrightarrow ROOOR \longrightarrow 2RO\cdot + O_2 \tag{64}$$

$$2ROOH$$
$$-2ROH$$

The presence of hydrogen on the carbon bearing the -OOH group
in secondary hydroperoxides results in the formation of carbonyl
compounds in addition to reactions 61-64 shown above when secondary
hydroperoxides are decomposed by catalytic quantities of transition
metal complexes (203). Primary alkyl hydroperoxides are decomposed
in the presence of metal complexes to give mainly alcohols, aldehydes
and in some cases carboxylic acids. For example cobalt(II)
octanoate catalyzes the decomposition of n-butylhydroperoxide in
pentane to give n-butyl alcohol (67%), oxygen (70%), and n-butyr-
aldehyde (32%) (209), eq. 65.

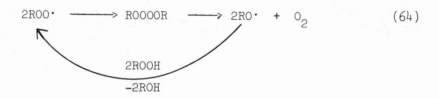

$$CH_3CH_2CH_2CH_2OOH \xrightarrow{\text{Co(Oct)}_2} CH_3CH_2CH_2CH_2OH + O_2 + CH_3CH_2CH_2CHO \tag{65}$$

Allylic hydroperoxides are likely intermediates during olefin
oxidation in the presence of many metal complexes. A recent
study by Arzoumanian, Blanc, Metzger and Vincent (210) has
elucidated the mechanism of the decomposition of cyclohexenyl
hydroperoxide in benzene catalyzed by [RhCl(PPH$_3$)$_3$], eq. 66.

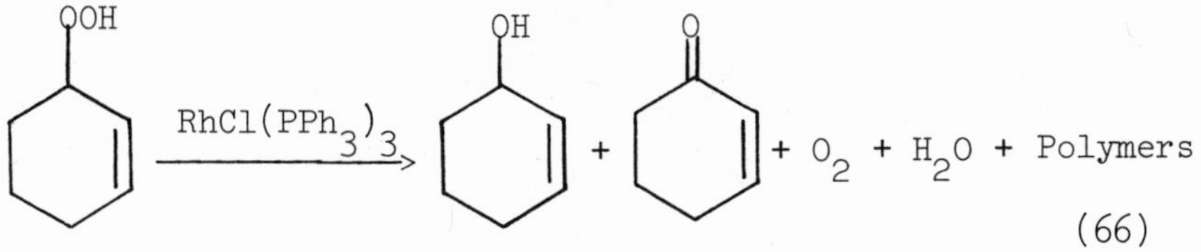

$$(66)$$

Inhibition studies showed that the process did not occur via a free radical chain mechanism but that each hydroperoxide molecule was catalytically decomposed to give one free radical. Oxygen was believed to be formed via the cyclohexeneylperoxy radical according to eq. 67.

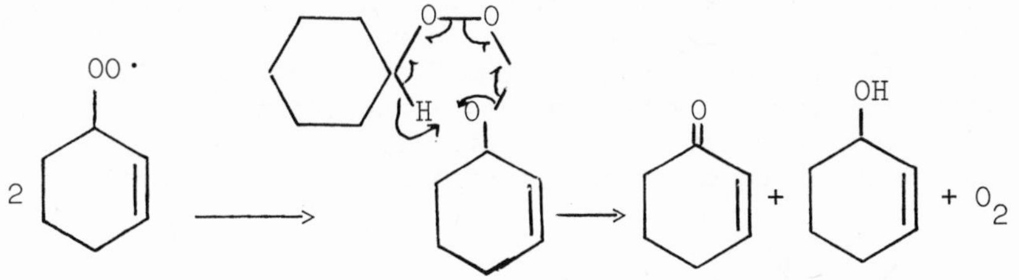

$$(67)$$

Kinetic studies are in accord with a monoelectronic Haber-Weiss mechanism involving Rh(I), and Rh(III) species, eq. 68-71.

$$\text{Rh(I)} + \text{ROOH} \longrightarrow \text{Rh(II)OH} + \text{RO}\cdot \qquad (68)$$

$$\text{Rh(II)OH} + \text{ROOH} \longrightarrow \text{Rh(I)} + \text{ROO}\cdot + \text{H}_2\text{O} \qquad (69)$$

$$\text{Rh(II)OH} + \text{ROOH} \longrightarrow \text{Rh(III)(OH)}_2 + \text{RO}\cdot \qquad (70)$$

$$\text{Rh(III)(OH)}_2 + \text{ROOH} \longrightarrow \text{Rh(II)OH} + \text{ROO}\cdot + \text{H}_2\text{O} \qquad (71)$$

In general, traces of hydroperoxides are difficult to remove from olefinic hydrocarbons. Their presence generates radicals according to eq. 62 - 64 when metals are added. The abstraction of allylic hydrogen atoms from an olefin is one of the most common radical chain pathways which results. Thus, radical initiated autoxidation gives rise to products of allylic oxidation, eq. 72, 73. Allylperoxy radicals can go on to stable oxidation products or can abstract hydrogen to give allylic hydroperoxides.

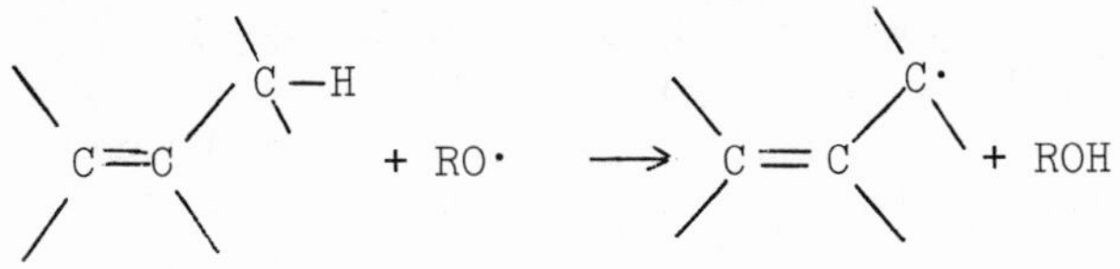

$$(72)$$

$$\ce{>C=C< + O2 -> >C=C<-C-OO.}$$ (73)

The nature of the metal catalyzed reactions of hydroperoxides with olefins can be divided into two general types. The reaction pathway which is preferred depends on the metal complex which is used as the catalyst. One group of reactions are characterized largely by homolytic bond cleavage resulting in products of free radical processes. Complexes which tend to promote this type of reaction are those of groups VIII and IB, in particular complexes of iron, cobalt and copper. The second group of reactions involve heterolysis of the O—O bond and result largely in the formation of an epoxide and an alcohol from reaction of the olefin with the hydroperoxide, eq. 74. Transition metal complexes which catalyze this reaction preferentially are those of groups IVB, VB and VIB, especially complexes of Mo, W, V, Cr and Ti.

$$\ce{ROOH + >C=C< ->[Catalyst] ROH + >C-C<}$$ (74)

Reaction 74 has found considerable utility in the commercial production of propylene oxide from propylene (212) and as a result has stimulated intense industrial interest and patent activity (212-214). The general scope and utility of this reaction has been the subject of several recent reviews. In contrast to metal catalyzed homolytic hydroperoxide decompositions, this reaction appears to clearly be an example of coordination catalysis. If the hydroperoxides initially formed during olefin autoxidation could be selectively reacted in this manner, selective catalytic reactions between dioxygen and olefins can be envisioned.

The properties of a metal complex which make it an efficient catalyst for the epoxidation of an olefin by a hydroperoxide have been considered by Sheldon and Van Doorn (220) and by Gould, Hiatt and Irwin (220a). Such a metal complex will have a high charge, a relatively small size and have low lying d orbitals which are at least partly unoccupied (220a). Complexes of metals in low oxidation states (e.g. $V(CO)_6$, $Mo(CO)_6$ or $W(CO)_6$ are rapidly oxidized by hydroperoxides to their highest oxidation states [Mo(VI), W(VI), V(V), Ti(IV)] which are the active catalysts (220). Since the major function of the catalyst is to withdraw electrons from the peroxidic oxygens, an active catalyst must be a good Lewis acid (220). It is also necessary that the

complex does not participate significantly in one electron transfer reactions under strongly oxidizing conditions (220a). Finally,
in order for the catalyst to be active it must form complexes
which are substitution labile (220a).

The Lewis acidity of the transition metal oxides increases
in the order CrO_3, MoO_3 >> WO_3 > TiO_2, V_2O_5 (220). Thus, it is
apparent why Mo(VI) is the most effective epoxidation catalyst.
Presumably Cr(VI) would also be expected to be a good catalyst,
however, Cr(VI) is a strong oxidant and readily causes decomposition of the hydroperoxide (220b). Group VIII complexes in many
instances exhibit the same difficulty.

The most important factor affecting the selectivity of the
epoxidation reaction, 74, is the choice of metal used as the
catalyst (219-222) and in general, molybdenum complexes are
superior catalysts for this reaction. Selectivity is also
dependent on olefin structure with the greater number of alkyl
groups on the double bond giving greater rates and yileds of
epoxides. Since molybdenum complexes have been most successfully
used in the epoxidation reaction, we will consider these catalysts
in greater detail.

On the basis of kinetic evidence, structure-reactivity
relationships, induction periods and inhibition by alcohols, a
schematic mechanism, eq. 75-78 for the molybdenum catalyzed
epoxidation of olefins has been put forward (219-223) which
accommodates most of the existing data.

$$Mo^n \longrightarrow Mo^{VI} \tag{75}$$

$$Mo^{VI} + ROOH \rightleftharpoons Mo^{VI}(ROOH) \tag{76}$$

$$Mo^{VI} + ROH \rightleftharpoons Mo^{VI}(ROH) \tag{77}$$

$$Mo^{VI}(ROOH) + C\ \ C \longrightarrow Mo^{VI} + ROH + \overset{C-C}{\underset{O}{\triangle}} \tag{78}$$

Sheldon has determined the nature of the catalytic Mo(VI),
eq. 75, species in a number of molybdenum catalyzed epoxidation
reactions (224). The active catalytic species in the epoxidation
of several olefins was found to be Mo(VI)-1,2-diol complexes
formed from reaction of the Mo(VI) complex with the epoxide product. Kaloustian, Lean and Metzger found that molybdic acid is
an intermediate in this sequence (225), eq. 79, 80. Sheldon
suggests that

$$MoL_x + ROOH \longrightarrow H_2MoO_4\ \ H_2O \tag{79}$$

$$\text{H}_2\text{MoO}_4 \ \text{H}_2\text{O} \ + \ 2 \ \triangle \quad \longrightarrow \quad \tag{80}$$

oxygen transfer, eq. 78, occurs from a hydroperoxide complex
of the Mo(VI)-1,2-diol system with protonation of the oxo group,
eq. 81.

$$\tag{81}$$

A completely different mechanism of epoxidation has
received support from some work which has appeared in the recent
literature. Mimoun, Seree de Roch and Sajus (30) observed that
the covalent Mo(VI) peroxo complexes $[\text{MoO(O}_2)_2\text{L}_x]$ reacted with
olefins to give epoxides in high yield, eq. 11. Since the
relative rates of olefin epoxidation are in the same order as
found for catalytic epoxidation (30, 226-229), eq. 11 has been
proposed as a possible mechanism for catalytic epoxidation.

Using labeling experiments, Sharpless and coworkers (230) have
found that the epoxidic oxygen arises exclusively from the
peroxo oxygen as expected from eq. 11. A recent report (231) of
the isolation of an active molybdenum diperoxo complex from an
epoxidation reaction mixture prompted the authors (231) to support
a mechanism (232), eq. 82, in which the actual epoxidizing agent
is a diperoxo complex formed in situ. Further support for this
hypothesis was reported recently by Arakawa and Ozaki (233) who
found that when the MoO_3 - catalyzed epoxidation of cyclohexene
was carried out in the presence of hexamethylphosphoramide, HMPA,
an intermediate diperoxo complex could be isolated. It is uncertain at this time to what extent this type, eq. 82, of
mechanism participates in olefin epoxidations.

The molybdenum and vanadium catalyzed epoxidation of olefins
exhibits considerable stereoselectivity. cis-Olefins give
cis-epoxides and trans-olefins give trans-epoxides (220), eq. 83.

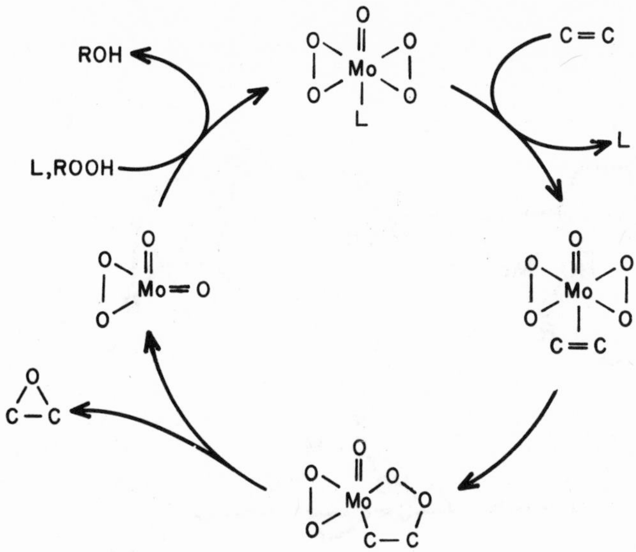

(82)

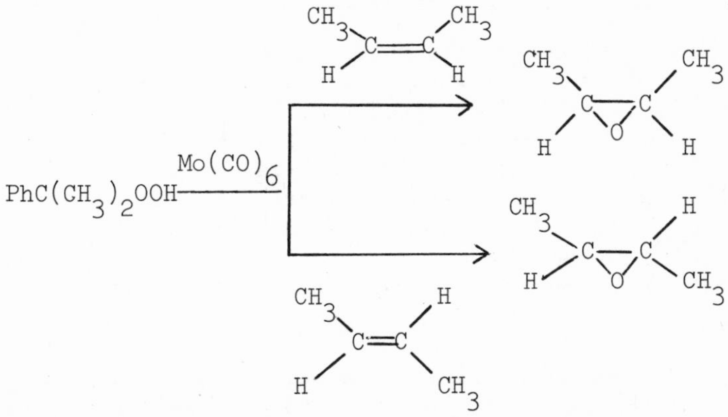

(83)

The stereochemistry of the epoxidation of cyclic olefins containing no complexing groups is determined solely by steric factors
(234,235). For example, the stereochemistry of the epoxidation
of α-pinene may be the result of attack by the hydroperoxide-
molybdenum complex from the least hinderered side, eq. 84.

When the olefin contains complexing groups, these groups usually
direct the steric course of the reaction. 2-Cyclohexene-1-ol,
for example, is epoxidized by hydroperoxides in the presence of
vanadium complexes to give cis-1,2-epoxycyclohexane-3-ol
(236-240), presumably via complexing of the hydroxyl group with

the metal center, eq. 85.

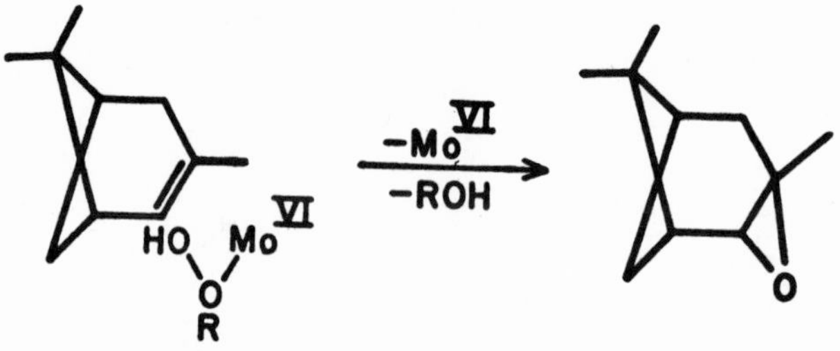

$$(84)$$

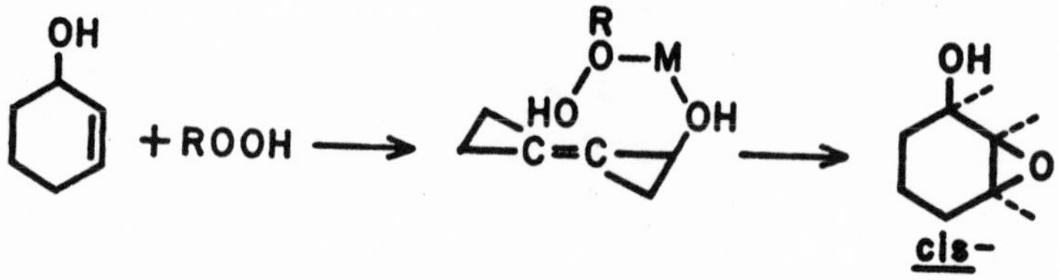

$$(85)$$

b) Reactions of Olefins with Dioxygen

It has long been known that metal salts and complexes promote the reaction of olefins with oxygen in the liquid phase.
Early work (241 and references cited therein) established that
during olefin oxidation in the presence of various copper, cobalt
and manganese salts, free radicals arise via decomposition of
a catalyst hydroperoxide complex formed from allylic hydroperoxide
generated in situ. Although the metal modifies the nature of the
observed products in many cases, most homogeneous metal-catalyzed
oxidations exhibit characteristics of free radical initiated
autoxidations.

Cyclohexene oxidations in the presence of a variety of acetylacetonates (242) were found to be free radical chain reactions
having the same homogeneous propagation steps and yielding as the
principle primary product, cyclohexenyl hydroperoxide. The metal
catalyzed decomposition of the primary product appeared to give
rise to varying amounts of the principle stable monomeric products
of oxidation: 2-cyclohexene-1-one, 2-cyclohexene-1-ol and
cyclohexene oxide.

The use of cobalt and manganese carboxylates to initiate the
oxidation of a large number of olefins such as the butenes
(243,244), propylene (245), oleic (246), linoleic (247), and
stearic (248,249) acids or their derivatives and α-methylstyrene

(250,251) is well known. The kinetics of oxidation of α-methyl-styrene in the presence of cobaltous and manganous acetylacetonates as well as copper phthalocyanine have been investigated (250,251). The results of this study led Kamiya to postulate a mechanism involving formation of radical species by a metal dioxygen complex, eq. 86, concurrent with radical generation by hydroperoxide composition.

$$MO_2 \ + \ \underset{/}{\overset{\backslash}{C}} = \underset{\backslash}{\overset{/}{C}} \ \longrightarrow \ MO_2 - \overset{|}{\underset{|}{C}} - \overset{|}{\underset{|}{C}} \cdot \qquad (86)$$

Although the predominant products of oxidations of olefins in the presence of cobalt complexes are usually α,β-unsaturated aldehydes and ketones, a report in the patent literature (252) asserts that cobalt di-(salicylal)-arylene-diimines, are efficient catalysts for the direct epoxidation of olefins with molecular oxygen in acetonitrile. A recent report by Budnik and Kochi (253) are shown that [Co(acac)$_3$] promotes the oxidation of olefins such as tert-butylethylene, norbornene, and 1,1-dineopentylethylene, which are incapable of undergoing allylic hydrogen abstraction. High yields of epoxides result, however, the reactions exhibit all the characteristics of free radical reactions. These authors propose the pathway shown in eqs. 87-90 to account for their observations.

Initiation: $Co(acac)_3 + O_2 \longrightarrow Co(oxide) + CO_2 + R\cdot$ (87)

Propagation: $R\cdot \ + \ O_2 \longrightarrow RO_2\dot{}$ (88)

(89)

(90)

Several years after the complexes: trans-[IrX(CO)(PPh$_3$)$_2$], (X = Cl, Br, I) were shown to exhibit reversible oxygen carrying properties, Collman, Kubota and Hosking found that compounds of this type as well as the rhodium(I) complex, [RhCl(Ph P)$_3$], prompted the oxidation of cyclohexene to give 2-cyclohexene-1-one predominantly. These authors postulated that 3-cyclohexene hydroperoxide was a likely intermediate and suggested that coordination of oxygen to the metal center was probably involved as a necessary

step in the oxidation reaction. It was then found (255) that the
oxidation of cyclohexene in the presence of [RhCl(Ph$_3$P)$_3$] exhibits
the characteristics of a radical chain process similar to reactions
carried out in the presence of cobalt carboxylates. It was postulated that hydroperoxides formed in situ were decomposed by
[RhCl(Ph$_3$P)$_3$] to radical species in a Haber-Weiss manner, eq. 62,
63 and that these radical species initiated autoxidation, eq.
64-74.

Since metal dioxygen complexes are usually diamagnetic, it
was of interest to determine whether the allylic hydroperoxide
formed in autoxidation reactions catalyzed by [RhCl(Ph$_3$P)$_3$]
arose via an "ene" addition pathway, eq. 91, or via conventional

$$(91)$$

radical pathways. To this end, Baldwin and Swallow studied the
oxidation of (+)-carvomenthene both photolytically and using
[RhCl(Ph$_3$P)$_3$] as the catalyst, (256). The photochemical reaction
gave the isomeric carvotanacetols by way of the corresponding
hydroperoxide. The rhodium catalyzed oxidation gave a mixture of
racemic carvotanacetone, piperitone, and alcohols. Since the
intermediate hydroperoxide was found to be optically stable under
reaction conditions, these authors conclude that the bulk of the
rhodium-catalyzed reaction proceeds via a symmetrized free radical
intermediate.

Fusi, Ugo, Fox, Pasini and Cenini (257) have investigated the
oxidation of cyclohexene in the presence of a number of metal complexes. It was found that d^8 complexes were more active catalysts
than d^{10} complexes. No relation was found between oxidation
activity and the strength of metal-dioxygen bonding. Activity
did not vary with the nature of the anionic ligand, X in complexes
of the type [M X(CO)(Ph$_3$P)$_3$]. A unique activity of [RhCl(Ph$_3$P)$_3$]
during early stages of reaction was noticed, however (257). For
example, minute traces of hydroperoxides totally eliminated the
pronounced induction periods which were observed when hydroperoxide
free cyclohexene was oxidized in the presence of [IrCl(CO)(Ph$_3$P)$_2$].
With [RhCl(Ph$_3$P)$_3$] as the catalyst, elimination of hydroperoxides
had little effect on initial reaction rates. It can be concluded
that in the case of the iridium-catalyzed reactions, radicals must
be generated from which allylic hydroperoxides can form; but
perhaps another pathway exists for hydroperoxide formation in the
presence of the rhodium complex.

James and Ochiai (258,259) studied the oxidation of a cyclo-octene rhodium(I) complex and found infrared evidence for cyclo-octene hydroperoxide as an intermediate. These authors postulated oxygen transfer directly to the olefin within the coordination sphere and suggested that such a scheme may account for the formation of 2-cyclohexene-1-one during oxidation of cyclohexene catalyzed by Rh(I) complexes.

Holland and Milner (260) re-examined this reaction recently in some detail. It was found that reaction of $[(C_8H_{14})_2RhCl]_2$ with dioxygen in benzene at 74°C gave equimolar amounts of 2-cyclooctene-1-one, cyclooctanone and water, eq. 92.

$$RhCl(L)(C_8H_{14})_2 \xrightarrow{O_2} RhCl(L)(C_8H_{14})_2(O_2) \xrightarrow{1/2\ O_2}$$

$$+ H_2O + \qquad\qquad\qquad\qquad (92)$$

Reactions proceeds via a pathway which is independent of radical chains and does not involve a Wacker cycle. By contrast, cyclo-octene was autoxidized by cobalt naphthenate by a radical pathway to give mainly cyclooctene oxide and this reaction was completely suppressed by the radical inhibitor 2,6-di-t-butyl-p-cresol. This inhibitor had no effect on reaction 92. Labeling experiments with added H_2O^{18} showed that water did not take part in this reaction, ruling out a Wacker process. DMA solutions of the analogous iridium(I) complex, $[IrCl(C_8H_{14})_2]_2$, absorb O_2 irreversibly with explusion of free cyclooctene to the solution (261). No evidence of any olefin oxidation products was found.

Dudley, Read and Walker (262) have reported a novel rhodium(I)-catalyzed oxidation of 1-olefins. Hexene-1, heptene-1 and octene-1 were converted to the corresponding methyl ketones with dioxygen at ambient temperature and pressure in benzene solutions of the complexes: $[RhH(CO)(PPh_3)_3]$, and $[RhCl(PPh_3)_3]$. Methyl ketones are not normally produced in Haber-Weiss initiated radical reactions. Furthermore, radical inhibitors such as hydroquinone or 2,6-di-t-butyl-p-cresol do not retard methyl ketone formation. Since these authors are unable to detect radical chain processes, they suggest that reactions involve co-oxygenation of coordinated PPh_3 and olefin at the metal center, eq. 93.

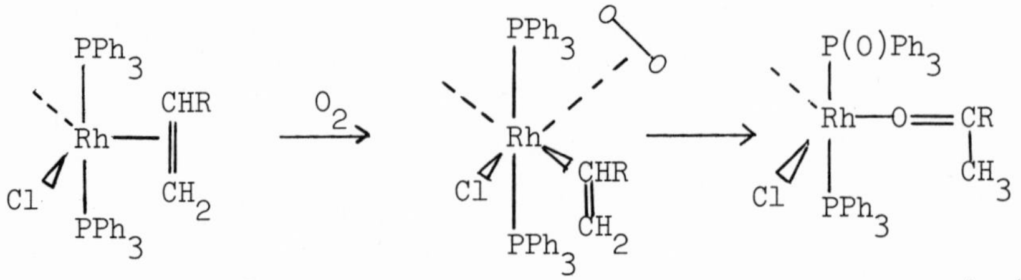

(93)

The oxidation of tetramethylethylene has also been studied
in the presence of the complexes: trans-$[MCl(CO)(Ph_3P)_2]$
(M = RH, Ir) (263). The oxidation was found to be rapid and quite
selective under mild conditions yielding 2,3-dimethyl-2,3-epoxy-
butane and 2,3-dimethyl-3-hydroxybutene-1 as the major oxidation
products. The reactions were inhibited by hydroquinone which
is consistent with a free radical initiated autoxidation. The
reaction of tetramethylethylene was more rapid than was oxidation
of less substituted olefins in the presence of the Rh(I) and Ir(I)
complexes suggesting that initial coordinative interaction between
the olefin and the metal center is not an important factor. An
allylic hydroperoxide was found to be a reaction intermediate
and eq. 93a suggested as the reaction pathway. Thus, an olefin
having allylic hydrogens capable of facile radical abstraction
to give a stable radical will generally prefer radical chain
autoxidation pathways.

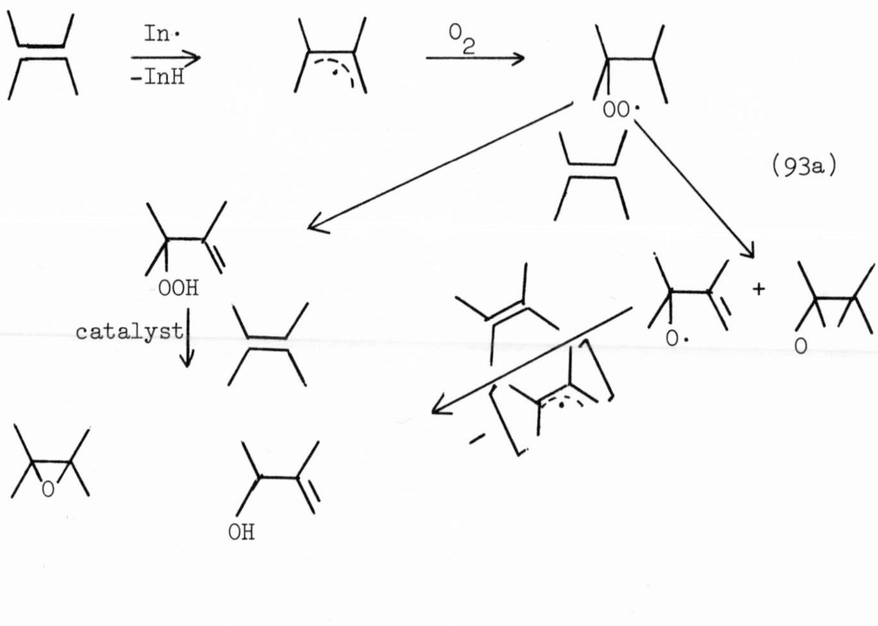

(93a)

The oxidation of styrene, an olefin having no allylic hydrogens capable of radical abstraction, has been the subject of extensive investigation (264-267), eq. 99. Reactions catalyzed by many metal complexes including $[IrCl(CO)(PPh_3)_2]$, $[RhCl(CO)(PPh_3)_2]$, and $[RhCl(PPh_3)_3]$ appear to be free radical in nature. A suggested reaction pathway, eqs. 94-98, is compatible with; a) radical inhibition studies; b) rapid decomposition

$$CH_2=CHPh + O_2 \longrightarrow [(CH\ CH(Ph)OO)]n \xrightarrow{M} PHCHO$$
$$+ CH_2O \tag{94}$$

$$CH_2O_2 + O_2 \xrightarrow{M} [HCO_3H] \tag{95}$$

$$[HCO_3H] + PhCH\ CH_2 \xrightarrow{M} PhCH\underset{O}{\triangle}CH_2 + HCOOH \tag{96}$$

$$HCOOH \xrightarrow{M} H_2 + CO_2 \tag{97}$$

$$H_2 + 1/2\ O_2 \xrightarrow{M} H_2O \tag{98}$$

of styrene polyperoxides by metal complex to give benzaldehyde and formaldehyde; c) observation of CO_2 and water as reaction by-products; and d) the observation that the yield of styrene oxide is approximately equal to the yield of benzaldehyde minus that of formaldehyde (267, 268).

Ruthenium and osmium complexes which most readily cleave the C=C double bond form the highest yields of styrene oxide during oxygenation at 75°C (267-269).

$$O_2 + PhCH = CH_2 \xrightarrow[75°C,\ 3.5\ Hrs]{(RuCl_2(PPh_3)_3)} PhCHO + CH_2O +$$
$$ 43\% \quad\quad 16\%$$

$$PhCH\underset{O}{-}CH_2 + PhCCH_3 \tag{99}$$
$$26\% 2\%$$

Cleavages of other olefins to give carboxylic acids and epoxides may proceed by a similar pathway (268), eq. 100.

$$CH_3(CH_2)_6CH = CH_2 + O_2 \xrightarrow{RuCl_2(PPh_3)_3} CH_3(CH_2)_6C\underset{O}{\overset{}{\triangle}}CH_2 +$$

$$CH_3(CH_2)_6COOH \qquad\qquad (100)$$

Products of the oxidative cleavage of olefinic double bonds are more prevalent during Rh(I)-catalyzed oxidation of cinnamaldehyde (269) and styryl acetate (269) than with styrene itself. Takao and co-workers (270) propose a mechanism involving a four-center intermediate having olefin and dioxygen in the coordination sphere for the Rh(I)-catalyzed cleavages.

Complexes of groups IVB, VB, and VIB taken as a class, tend to differ from complexes of VIIB, VIII, and IB regarding reactions to relevance to olefin oxidation. Group VIII complexes tend to react far more readily with molecular oxygen to form dioxygen complexes. Some group VIB dioxygen complexes, however, react smoothly with olefins to give epoxides while group VIII dioxygen complexes do not. Hydroperoxides are decomposed rapidly in the presence of many complexes of groups VIIB, VIII and IB whereas hydroperoxides are considerably more stable in the presence of many of the higher valent complexes of groups IVB, VB and VIB. Olefins are far more easily and selectively epoxidized in the presence of complexes of groups IVB, VB and VIB than in the presence of complexes of groups VIIB, VIII, and IB. Thus, it is not surprising that the product profile of oxidations of olefins using the two groups of complexes differs considerably in many instances.

Of these complexes, those most widely used in olefin oxidation have been complexes of molybdenum, vanadium, tungsten and chromium. Because of the well-known epoxidation activity of these complexes, a great effort has been made to selectively epoxidize propylene to the industrially important monomer, propylene oxide (271-283). Both the metal center and the ligand system seems to have an effect on reaction selectivity. Reaction conditions such as oxygen, partial pressure, and temperature as well as solvents are also important. It is generally accepted (271-283) that the group VB and VIB complexes serve to catalyze epoxidation of propylene with peroxidic double bond cleavage products as the epoxidizing agents, eq. 101-108.

$$CH_2 = CHCH_2O_2H \longrightarrow CH_3CHO + CH_2O \tag{101}$$

$$CH_2 = CHCH_2O_2^{\cdot} \longrightarrow CH_3\overset{\cdot}{C}O + CH_2O \tag{102}$$

$$CH_2CHO + O_2 \longrightarrow CH_3\overset{\cdot}{C}O + \cdot O_2H \tag{103}$$

$$CH_3\overset{\cdot}{C}O + O_2 \longrightarrow CH_3CO_3\cdot \tag{104}$$

$$CH_3CO_3\cdot + CH_2 = CHCH_2 \longrightarrow CH_3CO_2\cdot + CH_2 \overset{\triangle}{\underset{O}{\qquad}} CHCH_3 \tag{105}$$

$$CH_3CO_2\cdot + CH_2 = CHCH_3 \longrightarrow CH_3CO_2H + CH_2 = CHCH_2\cdot \tag{106}$$

$$CH_3CO_3\cdot + CH_2 = CHCH_3 \longrightarrow CH_3CO_3H + CH_2 = CHCH_2\cdot \tag{107}$$

$$CH_3CO_3H + CH_2 \ CHCH_3 \longrightarrow CH_3CO_2H + CH_2 \overset{}{\underset{O}{\qquad}} CH - CH_3 \tag{108}$$

As expected, minor reaction products are methanol, acetaldehyde, methyl formate, CO_2, CO, and polymeric residue. No allyl alcohol was detected (280).

The direct oxidation of a number of olefins other than propylene using Group V and VIB metal complexes have also been reported. Cyclohexene reacts with molecular oxygen in the presence of oxodiperoxo-bis(dimethylformamido) molybdenum(VI) to give cyclohexene oxide in 5-% selectivity at 12% olefin conversion (271). Ethylene may be oxidized with molecular oxygen to ethylene oxide in the presence of trialkoxychromium(III) complexes (284), hexene-1 to 1,2-epoxyhexane using the same complexes (284), and octene-1 to 1,2-epoxy octane using chromium acetylacetonate (285).

Whereas the major products of the $[C_5H_5Mo(CO)_3]$-catalyzed oxidation of substituted olefins are epoxides and allylic alcohols (236, 238), oxidation of substituted olefins in the presence of vanadium complexes gives rise to epoxy alcohols as the major products (236, 238, 286-289). When cyclohexene is the olefin used, reaction is observed to occur with a high degree of stereoselectivity (290), eq. 109.

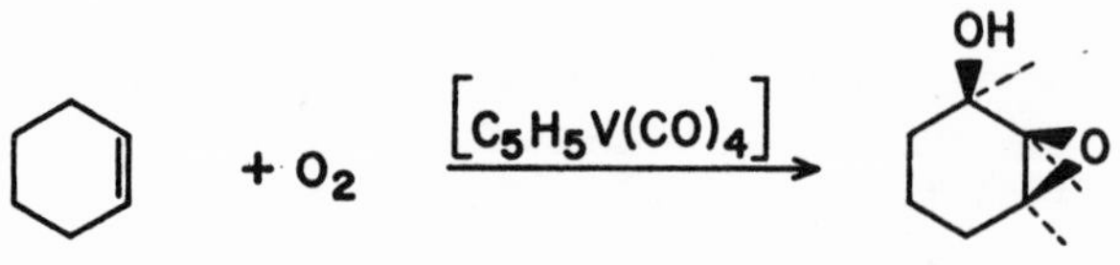

$$\text{cyclohexene} + O_2 \xrightarrow{\left[C_5H_5V(CO)_4\right]} \text{cis-1,2-epoxy-3-hydroxycyclohexane} \tag{109}$$

When cyclohexene, 5M in 1,2-dichloroethane, is oxidized in the presence of $10^{-2}M[C_5H_5V(CO)_4]$, <u>cis</u>-1,2-epoxy-3-hydroxycyclohexane is formed. At an olefin conversion of 10%, the yield of epoxy alcohol is 10% and the stereoselectivity is ~99%.

It has been noted (236, 238, 291) that the intermediate allylic hydroperoxide is stereoselectively converted to the <u>cis</u>-epoxy alcohol in the presence of vanadium complexes. Cross-product experiments (236, 238), eq. 110 and 111, experiments

$$(1) \quad \xymatrix{} + \text{OOH} \xrightarrow{\text{V-complex}} + \text{OH} \tag{110}$$

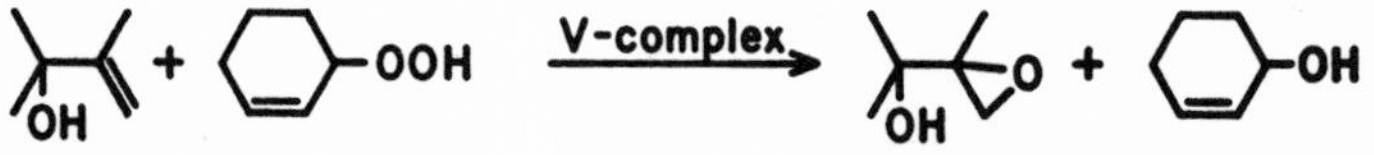

$$(2) \quad + \text{OH} \xrightarrow{\text{V-complex}} + \text{OH} \tag{111}$$

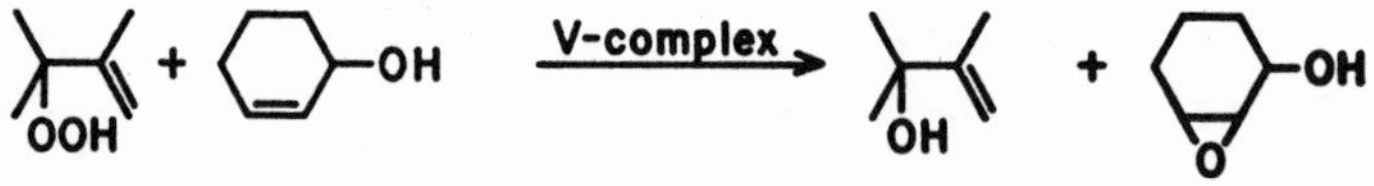

measuring relative rates of epoxidation of cyclohexene and 2-cyclo-hexene-1-ol (236,238), effects of added 2-cyclohexene-1-ol, and other data (236,238) indicate that in the case of vanadium-complex catalyzed oxidation of olefins, epoxy alcohols are formed <u>via</u> intermolecular epoxidation of allylic alcohols rather than by intramolecular rearrangement of allylic hydroperoxides, eq. 112. Thus vanadium complexes preferentially epoxidize small amounts of allylic oxygen species formed <u>in situ</u> to give epoxy alcohols whereas molybdenum complexes catalyze epoxidaton of the excess of unreacted olefin to give epoxides. The mechanism of vanadium catalyzed epoxidation of allylic alcohols has been discussed in an earlier section.

In summary, it appears that in most cases of olefin oxidation in the presence of salts or complexes of groups VII, VIII and IB, free radical pathways play a large role. In many cases, this is

(112)

the only pathway which occurs. The pathway depends both on the
nature of the olefin and that of the metal complex. Substituted
olefins having allylic hydrogens capable of easy abstraction
generally prefer radical pathways. First row metals generally
initiate radical pathways. Even in the absence of allylic
hydrogens radical pathways are possible and cobalt salts are involved in free radical initiated autoxidation. Rhodium complexes
appear to be capable of entering into some unusual reactions which
have been described as involving coordination of O_2 and olefin
in a catalytic cycle, however, iridium complexes do not react in
this manner. Ruthenium complexes are capable of catalyzing oxidative cleavage of double bonds and as a result give high yields
of spoxides as a co-product. The nature of this catalysis is
not clearly defined. It seems reasonable that coordination
catalysis is occurring in some of these systems and it is expected
that future work in this area will clarify the situation.
Molybdenum complexes selectively catalyze olefin epoxidation
reactions with hydroperoxides formed in situ while vanadium complexes give rise to epoxy alcohols.

CONCLUDING REMARKS

    Many examples of transition metal inhibited and mediated
radical autoxidations have found wide utility. On the other hand,
despite extensive efforts of many groups of workers for more than
a decade, efficient coordination catalysis of reactions of dioxygen
with organic substrates has been found in very few instances. These
cases have limited synthetic value and no commercial importance.
Even in cases wherein novel oxygen activation appears to play a
significant role, mechanisms are still poorly understood. Thus,
the dioxygen molecule has not been proven to be readily amenable
to activation through coordination by comparison with the more
easily manipulates small molecules such as H_2 or CO.

Clearly the dioxygen molecule is readily incorporated into
the coordination sphere of many metal complexes and is activated
by coordination. Paramagnetic dioxygen complexes, especially
the end-bonded type, react to form paramagnetic products, equations
52, 59, 74, giving rise to radicals which can initiate autoxidation. Competing radical autoxidations, however, are not the only
reason for failure to observe facile catalysis. Since side or
π-bonded peroxo complexes are diamagnetic, reactions with diamagnetic reactants to give diamagnetic products fulfill spin conservation requirements. Obviously, in many instances the product
of reaction between a metal dioxygen complex and a reactive substrate is a highly stable adduct and catalysis is impossible.
In other cases, once oxygen is transferred to a reactive substrate,
the metal is left in a state which does not permit entrance of a
second dioxygen molecule.

Examples of this can be seen both for paramagnetic side-
bonded dioxygen complexes which tend to be electrophilic in nature
(groups IVB-VIB) as well as for the nucleophilic type (group VIII).
Side-bonded dioxygen complexes such as [MoO_5(HMPA)] which prefer
to attack electron-rich olefins, form epoxides readily but leave
an MoO_3 moiety which does not react with O_2 to give the di-peroxo
MoO_5 group, eq. 113. Thus, reaction is only stoichiometric. A
nucleophilic paladium dioxygen complex reacts readily with electron-
deficient olefins to give a stable metalocycle which on thermal
decomposition leaves palladium in too high an oxidation state to
accept more oxygen (292), eq. 114. Again, catalysis does not occur.

$$\tag{113}$$

$$\tag{114}$$

Despite the problems which confront the chemist interested in pursuing coordination catalysis of dioxygen, investigations into this area can be most rewarding. Although radical pathways can predominate, formation of radicals under mild consitions can lead to selective control of autoxidation which would not be possible in the absence of the metal. The metal can direct the course of oxidation in new and interesting ways giving products which are not formed during classical autoxidation. The metal usually exerts its most profound effect during oxidation by catalyzing transformations of hydroperoxides formed in situ. Within the scope of these reactions lies the potential for selective catalytic synthesis of useful oxygen-containing monomers. Beyond this lies the possibility of finding new routes to oxidation products via coordination catalysis--but progress in this area may be slow. The challenge to chemists interested in this area is, however, to find the highly active metal complexes which will achieve the selective transformations whether they be by radical processes or via coordination catalysis.

REFERENCES

1. L. Vaska, Accounts of Chemical Research, 9, 175 (1976).
2. G. Henrici-Olivè and S. Olivè, Angew. Chem. Internat. Edit., 13, 29 (1974).
3. J. Valentine, Chem. Rev., 73, 235 (1973).
4. L. Klevan, J. Peone and S. Madan, J. Chem. Ed., 50, 670 (1973).
5. V. Choy and C. O'Connor. Coord. Chem. Rev., 9, 145 (1972/73).
6. A. Savitskii and V. Nelyubin, Russian Chemical Reviews, 44, 110, (1975).
7. M. Calvin, R. Bailes and W. Wilmarth, J. Amer. Chem. Soc., 68, 2254 (1946).
8. A. Crumbliss and F. Basolo, J. Amer. Chem. Soc., 92, 55 (1970).
9. J. Simplicio and R. Wilkins, J. Amer. Chem. Soc., 89, 6092 (1967).
10. C. Busetto, C. Neri, N. Palladino, and E. Perrotti, Inorg. Chim, Acta, 5, 129 (1971).
11. D. Diemente, B. Hoffman and F. Basolo, Chem. Commun., 467 (1970).
12. A. Sykes and J. Weil, Progr. Inorg. Chem., 13, 1 (1970).
13. J. Simplicio and R. Wilkins, J. Amer. Chem. Soc., 89, 6092 (1967).
14. F. Miller and R. Wilkins, J. Amer. Chem. Soc., 92, 2687 (1970).
15. F. Miller, J. Simplicio and R. Wilkins, J. Amer. Chem. Soc., 91, 1962 (1969).
16. N. Terry, E. Amma and L. Vaska, J. Amer. Chem. Soc., 94, 653 (1972).
17. B. Bosnich, W. Jackson, S. Lo, and J. McLaren, Inorg. Chem., 13, 2605 (1974).
18. L. Lindblom, W. Schaefer and R. Marsh, Acta Cryst., B27, 1461 (1971).

19. F. Fronczek and W. Schaefer, Inorg. Chim. Acta, 9, 143 (1974).

20. M. Bennett and R. Donaldson, J. Amer. Chem. Soc., 93, 3307 (1971).

21. S. Otsuka, A. Nakamura and Y. Tatsuno, Chem. Commun., 836 (1967).

22. S. Otsuka, A. Nakamura and Y. Tatsuno, J. Amer. Chem. Soc., 91, 6994 (1969).

23. S. Takahashi, S. Sonogashira and N. Hagihara, J. Chem. Soc., Jap., 87, 610 (1966).

24. G. Wilke, H. Schott and P. Heimbach, Angew. Chem. Int. Ed., 6, 92 (1967).

25. C. Cook and G. Jauhal, Inorg. Nucl. Chem. Lett. 3, 31 (1967).

26. J. Birk, J. Halpern, and A. Pickard, J. Amer. Chem. Soc., 90, 4491 (1968).

27. J. Halpern and A. Pickard, Inorg. Chem., 9, 2798 (1970).

28. H. Chan, Chem. Commun., 1550 (1970).

29. J. Baldwin, J. Swallow and H. Chan, Chem. Commun., 1407 (1971).

30. H. Mimoun, I. Seree De Roch and L. Sajus, Tetrahedron, 26, 37 (1970).

31. A. Johnson, J. Organometal. Chem., 49, C95 (1973).

32. J. Costa, A. Puxeddu, and L. Stefani, Inorg. Nucl. Chem. Letters, 6, 191 (1970).

33. G. McLendon and A. Martell, Coord. Chem. Revs., 19, 1 (1976).

34. D. Wagnerov, E. Schwertnerova, and I. VprekSiska, Coll. Czech. Chem. Commun., 38, 756 (1973).

35. A. Nishinaga, K. Watanabe, and T. Matsuura, Tetrahedron Lett., 1291 (1974).

36. A. Nishinaga, T. Togo and T. Matsuura, Chem. Commun., 896 (1974).

37. K. Garbett and R. Gillard, J. Chem. Soc. A, 1725 (1968).

38. J. Levinson and S. Robinson, J. Chem. Soc. A., 762 (1971).

39. T. Nappier and D. Meek, J. Amer. Chem. Soc., 94, 306 (1972).

40. M. Stiddard and R. Townsend, Chem. Commun., 1372 (1969).

41. K. Laing and W. Rober, Chem. Commun., 1556, 1558 (1968).

42. D. Christian, G. Clark, W. Roper, J. Waters and K. Whittle, Chem. Commun. 458 (1972).

43. P. Hayward, D. Blake, C. Nyman and G. Wilkinson, Chem. Commun., 987 (1967).

44. F. Cariati, R. Mason, G. Robertson and R. Ugo, Chem. Commun., 408 (1967).

45. P. Hayward and C. Nymon, J. Amer. Chem. Soc., 93, 617 (1971).

46. R. Ugo, F. Conti, S. Cenini, R. Mason and G. Robertson, Chem. Commun., 1498 (1968).

47. P. Hayward, S. Saftich and C. Nyman, Inorg. Chem., 10, 1311 (1971).

48. K. Sharpless, J. Townsend and D. Williams, J. Amer. Chem. Soc., 94, 295 (1972).

49. S. Otsuka, A. Nakamura, Y. Tatsuo, and M. Miki, J. Amer. Chem. Soc., 94, 3761 (1972).

50. C. Brown and G. Wilkinson, Chem. Commun., 70 (1971).

51. C. Brown, D. Georgiou, and G. Wilkinson, J. Chem. Soc., A, 3120 (1971).

52. W. Siegl, S. Lapporte and J. Collman, Inorg. Chem., 10, 2158 (1971).

53. Y. Iwashita and A. Hayata, J. Amer. Chem. Soc., 91, 2525 (1969).

54. K. Grundy, K. Laing and W. Rober, Chem. Commun., 1500 (1970).

55. L. Johnson and J. Page, Can. J. Chem., 47, 4241 (1969).

56. C. Giannotti, A. Gaudenier, and C. Fontaine, Tetrahedron Lett., 37, 3209 (1970).

57. B. Booth, R. Hazeldine and G. Neuss, Chem. Commun., 1074 (1972).

58. L. Harvie and F. McQuillan, Chem. Commun., 369 (1976).

59. J. Birk, J. Halpern, and A. Pickard, Inorg. Chem., 7, 2672 (1968).

60. S. Takahashi, K. Sonogashira and N. Hagihara, Nippon Kagaku Zasshi, 87, 610 (1966).

61. J. Halpern (private communication).

62. E. Stern, "Transition Metals in Homogeneous Catalysis," Marcel Dekker, Inc., New York, p. 138 (1971).

63. B. Graham, K. Laing, C. O'Connor and W. Roper, Chem. Commun., 1272 (1970).

64. B. Graham, K. Laing, C. O'Connor and W. Roper, J. Chem. Soc. Dalton, 1237 (1972).

65. T. Kahn, R. Andal and P. Manoharan, Chem. Commun., 561 (1971).

66. S. Cenini, A. Gusi and G. Capparella, J. Inorg. Nucl. Chem., 33, 3576 (1971).

67. B. James, L. Markham, A. Rattray and D. Wang, Inorg. Chim. Acta, 20L25 (1976).

68. R. Poddar and U. Agarwala, Inorg. Nucl. Chem. Letters, 9, 785 (1973).

69. B. VanVugt, N. Koole, W. Drenth and F. Kuijpers, Rec. Trav. Chim., 92, 1321 (1973).

70. R. Augustine and J. van Peppen, Chem. Commun., 497 (1970).

71. C. Dudley and G. Read, Tetrahedron Lett., 5273 (1972).

72. H. Arai and J. Halpern, Chem. Commun., 1571 (1971).

73. K. Takao, Y. Fujiwara, T. Imanaka and S. Teranishi, Bull. Chem. Soc. Japan, 43, 1153 (1970).

74. D. Schmidt and J. Yoke, J. Amer. Chem. Soc., 13, 637 (1971).

75. J. Drapier and A. Hubert, J. Organometal. Chem., 64, 385 (1974).

76. J. Halpern, B. Goodall, G. Khare, H. Lim and J. Pluth, J. Amer. Chem. Soc., 97, 2301 (1975).

77. R. Barral, C. Bocard, I. Seree deRoch and L. Sajus, Tetrahedron Lett. 1693 (1972).

78. R. Barral, C. Bocard, I. Seree deRoch and L. Sajus, Fr., 2, 115,598 (1972).

79. R. Barral, C. Bocard, I. Seree deRoch and L. Sajus,
     Kinet. Katal., 14, 164 (1973).
80. F. Moore, M. Larson, Inorg. Chem., 6, 998 (1967).
81. N. Sutin and J. Yandell, J. Amer. Chem. Soc., 95, 4847 (1973).
82. L. Avdeeva and A. Mashkina, Neftekhimiya, 14, 461 (1974).
83. M. Ledlie and I. Howell, Tetrahedron Lett., 785 (1976).
84. J. Trocha-Grimshaw and H. Henbest, Chem. Commun., 1035 (1968).
85. H. Henbest and J. Rocha-Grimshaw, J. Chem. Soc., Perkin Trans.,
     1, 607 (1974).
86. N. Connon, Fr., 2,095,718 (1972).
87. N. Connon, Org. Chem. Bull., 44, 1 (1972).
88. I. Seree de Roch, Fr., 1, 540,284 (1968).
89. L. Kuhnen, Angew. Chem., Int. Ed., 5, 893 (1966).
90. V. List and L. Kuhnen, Erdol und Kohle, 20, 192 (1967).
91. M. Weininger, I. Taylor and E. Amma, Chem. Commun., 1172 (1971).
92. W. Brinigar, C. Chang and T. Traylor, J. Amer. Chem. Soc.,
     96, 5597 (1974).
93. S. Otsuka and M. Tatsuno, Japan, 70 19,884 (1970).
94. E. Derouane, J. Braham and R. Hubin, J. Catal., 35, 196 (1974).
95. A. Terentev and Y. Mogilyansky, Dokl. Akad. Nauk SSSR, 103,
     91 (1955); Chem. Abs., 50, 4807 (1956).
96. K. Kinoshita, Bull. Chem. Soc. Japan, 32, 78 (1956).
97. K. Kinoshita, Bull. Chem. Soc. Japan, 32, 777 (1956).
98. A. Terentev and Y. Mogilyansky, J. Gen. Chem. USSR, 31,
     298 (1961).
99. K. Pausacker, J. Chem. Soc., 1989 (1953).
100. K. Pausacker, et al., J. Chem. Soc., 4003 (1954).
101. K. Wurthrich and S. Fallab, Helv. Chim. Acta., 47, 1440 (1970).
102. V. Van Rheenen, Chem. Commun., 314 (1969).
103. T. Ho, Synth. Commun., 4, 135 (1974).
104. T. Itoh, K. Kaneda, I. Watanabe, S. Ikeda and S. Teranishi,
     Chem. Lett., 227 (1976).
105. E. Balogh-Hergovich and G. Speier, Reaction Kinetics and
     Catalysis Letters, 3, 139 (1975).
106. M. Barker and S. Perumal, Tetrahedron Lett., 349 (1976).
107. H. Bach, U.S. 3, 719,701 (1973)
108. H. Takahashi, T. Kajimoto and J. Tsuji, Synth. Commun., 2,
     181 (1972).
109. J. Tsuji, H. Takahashi and T. Kajimoto, Tetrahedron Lett.,
     4573 (1973).
110. C. Kramer, G. Davies, R. Davis, and R. Slaven, Chem. Commun.,
     606 (1975).
111. D. Wagnerova, E. Schwertnerova and J. Veprek-Siska, Coll.
     Czech Commun., 38, 756 (1973).
112. L. Dohnal and J. Zyka, Microchem., J., 19, 63 (1974);
     Chem. Abs., 80, 108091w
112a. M. N. Dufour-Ricroch and A. Gaudemer, Tetrahedron Letters,
     4079 (1976).
112b. A. Nishinaga, Chem. Lett., 273, (1975).

113. K. Maeda, I. Moritani, T. Hosokawa and S. Murahashi, Tetrahedron Lett., 797 (1974).
114. L. Kuhnen, Chem. Ber., 99, 3384 (1966).
115. M. Sheng and J. Zajacek, J. Org. Chem., 33, 558 (1968).
116. G. Tolstikov, U. Dzhemilev, V. Yur'ev, A. Pozdeeva and F. Gerchikova, Zh. Obshchei Khim., 43, 1360 (1973).
117. G. Tolstikov, U. Jemilev, V. Jur'ev, F. Gershanov and S. Rafikov, Tetrahedron Lett., 2807 (1971).
118. G. Tolstikov, U. Dzhemilev, V. Yur'ev, Zh. Obshchei Khim., 8, 2204 (1972).
119. K. Kosswig, Liebigs Ann. Chem., 749, 206 (1971).
120. G. Howe and R. Hiatt, J. Org. Chem., 35, 4407 (1970).
121. H. DeLaMare, J. Org. Chem., 25, 2114 (1960).
122. Halcon Int. Inc., Belg. Patents 661,500 (1965) and 668,811 (1975).
123. N. Johnson and E. Gould, J. Amer. Chem. Soc., 95, 5198 (1973).
124. N. Johnson and E. Gould, J. Org. Chem., 39, 407 (1974).
125. J. Kiji and J. Furukawa, Chem. Commun., 977 (1970).
126. G. Mercer, J. Shu, T. Rauchfuss and D. Roundhill, J. Amer. Chem. Soc., 97, 1967 (1975).
127. J. Byerly and J. Lee, Can. J. Chem., 45, 3025 (1967).
128. R. McAndrew and E. Peters, Can. Met. Quart., 3, 153 (1964).
129. S. Nakamura and J. Halpern, J. Amer. Chem. Soc., 83, 4102 (1961).
130. C. Costa, G. Mestroni, G. Pellizer and T. Licari, Inorg. Nucl. Chem. Letters, 5, 515 (1969).
131. E. Hirsch and E. Peters, Can. Met. Quart., 3, 137 (1964).
132. A. C. Harkness and J. Halpern, J. Amer. Chem. Soc., 83, 1258 (1961).
133. B. R. James and G. L. Rempel, Chem. Commun., 158 (1967).
134. B. R. James and G. L. Rempel, J. Chem. Soc. (A), 78 (1969).
135. G. Rosenberg, Ph.D. Thesis, University of British Columbia (1974).
136. J. Stanko, G. Petrov and C. Thomas, Chem. Commun., 1100 (1969).
137. J. Bayston and M. Winfield, J. Catal., 9, 217 (1967).
138. L. Lee and G. Schrauzer, J. Amer. Chem. Soc., 90, 5274 (1968).
139. J. Bercaw, L. Goh and J. Halpern, J. Amer. Chem. Soc., 94, 6535 (1972).
140. T. Kruck and M. Noach, Chem. Ber., 97, 1693 (1964).
141. W. Friedrich, Z. Naturforsch, B., 25, 1431 (1970).
142. J. Nicholson, J. Powell and B. Shaw, Chem. Commun., 174 (1966).
143. J. Powell and B. Shaw, J. Chem. Soc. (A), 583 (1968).
144. W. Dent, R. Long and A. Wilkinson, J. Chem. Soc., 1585 (1964).
145. V. Likholobov, V. Zudin, N. Eremenko, and Y. Ermakov, Kinet. Katal., 15, 1613 (1974).
146. W. Lloyd and D. Rowe, U.S. 3,849,336 (1975).

147. T. Imanaka, S. Matsumoto, S. Wakimura and S. Teranishi,
      Kogyo Kagaku Zasshi, 74, 1071 (1971); Chem. Abs., 75,
      67955g (1972).

149. T. Imanaka, S. Matsumoto, S. Wakimura and S. Teranishi,
      Kogyo Kagaku Zasshi, 74, 2222 (1971).

151. V. Avdeev, L. Kozhevina, K. Matveev, I. Ovsyannikova,
      N. Eremenko and L. Rachkovskaya, Kinet. Katal., 15, 935
      (1974).

152. D. Fenton and P. Steinwand, J. Org. Chem., 39, 701 (1974).

153. W. Gaenzler, K. Klaus and G. Schroeder, Ger. Offen, 2,213,435
      (1973).

154. H. Verter, "The Chemistry of the Carbonyl Group," Chapter 2,
      Vol. 2, Interscience Publishers, New York (1970).

155. J. McNesby and C. Heller, Jr., Chem. Rev., 54, 325 (1954).

156. C. Bawn and J. Jolley, Proc. Roy. Soc. (London), A237,
      297 (1956).

157. F. Marta, E. Boga and M. Matok. Discuss. Faraday Soc., 46,
      173 (1968).

158. A. Ivanov, T. Grimalovskaya and L. Ivanova, Zh. Fiz. Khim.,
      49, 893 (1975); Chem. Abstr., 83, 42967q (1975).

159. N. Digurov and V. Sedlyarov, Tr. Mosk. Khim-Technol. Inst.,
      66, 76 (1970; Chem. Abstr., 75, 76334f (1971).

160. S. Imamura and Y. Takegami, Kogyo Kagaku Zasshi, 74, 2490
      (1971); Chem. Abstr., 76, 45483v (1972).

160a. T. Matsuzaki, J. Imamura and N. Ohta, Kogyo Kagaku Zasshi,
      71, 706 (1968).

161. C. Bawn, F. Hobin and L. Raphael, Proc. Roy. Soc. (London),
      A237, 297 (1956).

162. E. Hesse and H. Steger, Ger. (East), 60,553 (1968).

163. M. Tezuka, O. Sekiguchi, Y. Ohkatsu, and T. Osa, Bull. Chem.
      Soc. Japan, 49, 2765 (1976).

164. H. Jun-Ichi, S. Yuasa, N. Yamazo, I Mochida, and T. Seiyama,
      J. Catal., 36, 93 (1975).

165. V. Komissarov and E. Denisov, Zh. Fiz. Khim., 43, 769 (1969).

166. V. Komissarov and E. Denisov, Zh. Fiz. Khim., 44, 390 (1970).

167. V. Komissarov and E. Denisov, Neftekhimiya, 8, 595 (1968).

168. V. Komissarov and E. Denisov, Neftekhimiya, 7, 420 (1967).

169. M. Saitova and V. Komissarov, Kinet. Katal., 13, 496 (1972).

170. R. Kozlenkova, V. Kamzolkim and A. Bashkirov, Neftekhimiya,
      10, 707 (1970).

171. H. Den Hertog, Jr., and E. Kooyman, J. Catal., 6, 357 (1966).

172. S. Suzuki, T. Tokumaru, E. Ando and Y. Watanabe, Japan Kokai,
      74 00,235 (1974).

173. P. Sukhopar, B. Zubko and K. Chervinskii, Zh. Prikl. Khim.,
      47, 1155, (1974).

174. S. Kamath and S. Chandalia, J. Appl. Chem. Bio-Technol, 23,
      469 (1973).

175. E. Baranova and K. Chervinskii, Khim. Technol., 19, 151 (1971).
176. K. Chervinskii and V. Mal'tsev, Neftekhimiya, 7, 264 (1967).
177. N. Sakai, M. Ogawa and M. Kitabatake, Japan, 69 12,128 (1969).
178. M. Kusunoki and M. Ogawa, Japan, 69 05,858 (1969).
179. M. Ogawa, M. Kusunoki and M. Kitabatake, Japan, 69 26,283 (1969).
180. R. Barker, et al., U.S. 3,234,271 (1968).
181. E. Yasui, et al., Brit. 1,169,777 (1969).
182. R. Lidov, U.S. 3,361,806 (1968).
183. Y. Ishimoto, H. Togawa and S. Nakahachi, Japan, 72 26,768 (1972).
184. H. Charman, Brit. 1,114,885 (1968).
185. K. Takagi and T. Ishida, Ger. Offen., 2,124,712 (1971).
186. Ashai Chemical Industry Co., Brit. 1,103,885 (1968).
187. I. Korsak, V. Agabekov, and N. Mitskevich, Neftekhimiya, 15, 130 (1975).
188. Y. Kamiya, Kogyo Kagaku Zasshi, 74, 1811 (1971).
189. A. Semenchenko, V. Solyanikov and E. Denisov, Kinet. Katal., 13, 1153 (1972).
190. C. Gardner, G. Gilbert and W. Morris, Brit. 1,250,192 (1971).
191. W. Morris, Ger. Offen., 1,912,878 (1969).
192. W. Morris, Ger. Offen., 2,037,189 (1971).
193. P. Camerman and J. Hanotier, Fr., 2,094,808 (1972).
194. C. Cullis and A. Fish, "Chemistry of the Carbonyl Group, Chapter 2, Carbonyl-Forming Oxidations," s. Patai, Ed., John Wiley and Sons, New York, 1966.
195. W. Brackman, U.S. 2,883,426 (1959).
196. P. Camerman and J. Hanotier, Fr., 2,095,160 (1972).
197. S. Matsuda, A. Uchida and T. Yamazi, Nippon Kagaku Kaishi, 296 (1973).
198. V. Sapunov, E. Selyutina, O. Tolchinskaya and N. Libedev, Kinet. Katal., 15, 605 (1974).
199. A. Savitskii, Zh. Obsch. Khim., 44, 1548 (1974).
200. J. Groves and M. Van Der Puy, J. Amer. Chem. Soc., 96, 5274 (1974).
201. J. Groves and M. Van Der Puy, Tetrahedron Lett., 1949 (1975).
202. J. Groves and M. Van Der Puy, J. Amer. Chem. Soc., 97, 7118 (1975).
203. G. Sosnovsky and D. Rawlinson, "Organic Peroxides," Vol. II, Ed D. Swern, John Wiley and Sons, Inc., New York, p. 153, 1971.
204. D. Lindsay, J. Howard, E. Horswill, K. Ingold and T. Cobbley, Can. J. Chem., 51, 870 (1973).
205. J. Bennett and J. Howard, J. Amer. Chem. Soc., 95, 4008 (1973).
206. A. Factor, C. Russel and T. Traylor, J. Amer. Chem. Soc., 87, 3692 (1965).
207. R. Hiatt and T. Traylor, J. Amer. Chem. Soc., 87, 3766 (1965).
208. W. Pritzkow and K. Muller, Ber., 89, 2321 (1956).

209. R. Hiatt, K. Irwin and C. Gould, J. Org. Chem., _33_, 1430 (1968).

210. H. Arzoumanian, A. Blanc, J. Metzger and J. Vincent, J. Organometal. Chem., _82_, 261 (1974).

211. H. Arzoumanian, A. Blanc, J. Metzger and J. Vincent, J. Organometal. Chem., _82_, 261 (1974).

212. J. Kollar, U.S. 3,350,422 (1967: U.S. 3,351,635 (1967); U.S. 3,360,584 (1967).

213. N. Sheng and J. Zajacek, Can., 779,502 (1968); Can., 799,503 (1968); Can., 799,504 (1968).

214. I. DeRoch and P. Menguy, Fr., 1,505,337 (1967); Fr., 1,505,332 (1967).

215. R. Hiatt in "Oxidation Techniques and Applications in Organic Synthesis," (R. Augustine, Ed.), Vol. 2, pp. 113-138, Marcel Dekker, Inc., New York, 1971.

216. D. Metelitsa, Russian Chemical Reviews (Engl. Transl.), _41_, 807 (1972).

217. F. Mashion and S. Kaito, Yuki Gosei Kagaku Kyokaishii, _26_, 367 (1968).

218. A. Doumaux, "Oxidation" Vol. 2, Ed., R. Augustine, Marcel Dekker, Inc., New York, 141-185, 1971.

219. N. Indicator and W. Brill, J. Org. Chem., _30_, 2074 (1964).

220. M. Sheng and J. Zajecek, Advan. Chem. Ser., _76_, 418 (1968).

220a. E. Gould, R. Hiatt and K. Irwin, J. Amer. Chem. Soc., _90_, 4573 (1968).

220b. T. Jochsberger, D. Miller, F. Herman and N. Indictor, J. Org. Chem., _36_, 4078 (1971).

221. R. Sheldon and J. Van Doorn, J. Catal, _31_, 427 (1973).

222. M. Sheng, J. Zajacek and T. Baker, Symposium on New Olefin Chemistry, Houston, Texas, February 1970.

223. G. Howe and R. Hiatt, J. Org. Chem., _36_, 2493 (1971).

224. R. Sheldon, Recl. Trav. Chim. Pays-Bas, _92_, 367 (1973).

225. J. Kaloustian, L. Lena and J. Metzger, Tetrahedron Lett., 599 (1975).

226. H. Mimoun, I. Seree DeRoch, L. Sajus and P. Menguy, Fr., 1,549,184 (1968).

227. H. Mimoun, I. Seree DeRoch, P. Menguy, and L. Sajus, Ger. Offen., 1,815,998 (1969).

228. H. Mimoun, I. Seree DeRoch, P. Menguy and L. Sajus, Ger. Offen., 1,817,717 (1970).

229. H. Arakawa, Y. Moro-oka and A. Ozaki, Bull. Chem. Soc. Japan, _47_, 2958 (1974).

230. K. Sharpless, J. Townsend and D. Willeams, J. Amer. Chem. Soc., _94_, 296 (1972).

231. K. Khcheyan, L. Samter and A. Sokolov, Neftekhimiya, _15_, 415 (1975).

232. J. Kaloustian, L. Lena and J. Metzger, Bull. Socl. Chim. France, 4415 (1971).

233. H. Arakawa and A. Ozaki, Chem. Lett., 1245 (1975).
234. V. Yur'ev, I. Gailyunas, Z. Isaeva and G. Tolstikov, Izv.
      Akad. Nauk SSSR, Ser. Khim., 919 (1974).
235. V. Yur'ev, I. Gailyunas, L. Spirikhin and G. Tolstikov,
      Shur. Obshch. Khim., 45, 2312 (1975).
236. J. Lyons, Homogeneous Catalysis-II, Joint Symposium of the
      Division of Industrial and Engineering Chemistry and
      Petroleum Chemistry, 166th Meeting, ACS. Chicago, Illinois,
      August (1973).
237. K. Sharpless and R. Michaelson, J. Amer. Chem. Soc., 95,
      6136 (1973).
238. J. Lyons, "Catalysis in Organic Synthesis," P. Rylander
      and H. Greenfield, Ed., Academic Press, New York, pp 235-
      255 (1976).
239. Y. Paushkin, I. Kolesnikov, B. Sherbanenko, S. Nizova, and
      L. Vilenskii, Kinet. Katal., 13, 493 (1972).
240. S. Tanaka, H. Yamamoto, H. Nozaki, K. Sharpless,
      R. Michaelson and J. Cutting, J. Amer. Chem. Soc., 96, 5254
      (1974).
241. A. Chalk and J. Smith, Trans. Faraday Soc., 53, 1214 (1957).
242. E. Gould and M. Rado, J. Catal., 13, 238 (1969).
243. G. Yakimova and O. Levanevskii, Izv. Akad. Nauk Kirg. SSR,
      68, (1972); Chem. Abstr., 77, 151379w (1972).
244. J. Imamura, T. Saito, and N. Ohta, Kogyo Kaguku Zasshi, 71,
      1642 (1968).
245. M. Naylor, U.S. 3,271,447 (1966).
246. S. Nan'ya and K. Fuduzumi, Kogyo Kagaku Zasshi, 72, 589
      (1969).
247. T. Labuza, J. Maloney and M. Karel, J. Food Sci., 31, 885
      (1969).
248. A. Schwab, J. Amer. Oil Chem. Soc., 50, 74 (1973).
249. A. Schwab, E. Frankel, E. Dufek and J. Cowan, J. Amer. Oil
      Chem. Soc., 49, 75 (1972).
250. Y. Kamiya, J. Catal., 24, 69 (1972).
251. Y. Kamiya, Tetrahedron Lett., 4965 (1968).
252. H. Klein, C. Bell and J. Coyle, Canadian Patent, 888,221
      (1971).
253. R. Budnik and J. Kochi, J. Org. Chem., 41, 1384 (1976).
254. J. Collman, M. Kubota and J. Hosking, J. Amer. Chem. Soc.,
      89, 4809 (1967).
255. V. Kurkov, J. Pasky and J. Lavigne, J. Amer. Chem. Soc.,
      90, 4744 (1968).
256. J. Baldwin and J. Swallow, Angew Chem., Int. Ed., 8, 601
      (1969).
257. A. Fusi, R. Ugo, F. Fox, A. Pasini and S. Cenini, J.
      Organometal. Chem., 26, 417 (1971).
258. B. James and E. Ochiai, Can. J. Chem., 49, 975 (1971).
259. B. James and F. Ng., Chem. Commun., 908 (1970).

260. D. Holland and D. Milner, J. Chem. Soc. Dalton, 2440 (1975).
261. C. Chan and B. James, Inorg. Nucl. Chem. Letters, 9, 135 (1973).
262. K. Allison, M. Chambers and G. Foster, Brit. Pat., 1,206,166 (1970).
263. J. Lyons and J. Turner, J. Org. Chem., 37, 2882 (1972).
264. K. Takao, M. Wayaku, Y. Fujiwara, T. Imanaka and S. Teranishi, 43, 3898 (1970).
265. J. Farrar, D. Holland and D. Milner, J. Chem. Soc. Dalton, 815 (1975).
266. J. Lyons and J. Turner, Tetrahedron Letters, 2903 (1972).
267. J. Lyons, Adv. Chem. Ser., 132, 64 (1974).
268. M. Pudel and Z. Maizus, Izv. Akad. Nauk SSSR., Ser Khim., 43 (1975).
269. K. Takao, M. Wayaku, Y. Fujiwara, T. Imanaka and S. Teranishi, Bull. Chem. Soc. Japan, 45, 1505 (1972).
270. K. Takao, H. Azuma, Y. Fujiwara, T. Imanaka, and S. Teranishi, Bull. Chem. Soc. Japan, 45, 2003 (1972).
271. C. Bocard, C. Gadelle, H. Mimoun and I. Seree deRoch, Fr., 2,044,007 (1971).
272. J. Alagy, C. Busson, C. Gadelle, I. Seree deRoch, and L. Sajus, Fr. 2,052,068 (1971).
273. J. Rouchaud and J. Mawaka, J. Catal., 19, 172 (1970).
274. J. Rouchaud and M. DePauw, Bull. Soc. Chim. Fr., 2905, (1970), ibid, 2914 (1970).
275. J. Rouchaud and A. Mumbieni, Bull. Soc. Chim. Fr., 2907 (1970).
276. J. Rouchaud and F. Mingiedi, Bull. Soc. Chim. Fr., 2909 (1970) ibid, 2912 (1970).
277. J. Rouchaud and F. Mingiedi, Bull. Chim. Soc. Belg., 78, 285 (1969).
278. J. Rouchaud and P. Nsumba, Bull. Chim. Soc. Belg., 77, 551 (1968).
279. J. Rouchaud, Bull. Soc. Chim. Fr., 1189 (1971).
280. E. deRuiter, Erdol und Kohle, 510 (1972).
281. S. Cavitt, U.S., 3,856,826 (1974).
282. S. Cavitt, U.S., 3,856,827 (1974).
283. Ger. Offen., 2,313,023 (1973).
284. V. Tulupov and T. Zakhar'eva, Zh. Fiz. Khim., 49, 272 (1975).
285. P. Hayden, Brit. 1,209,321 (1970).
286. K. Allison, G. Foster, P. Johnson and M. Sparke, presented at the Amer. Chem. Soc. National Meeting, Detroit (1965).
287. K. Allison, P. Johnson, G. Foster and M. Sparke, Ind. Eng. Chem., Prod. Res. Develop., 5, 166 (1966).
288. K. Allison, Belg., 640,204 (1964).
289. K. Allison, U.S. 3,505,360 (1970).
290. J. Lyons, Tetrahedron Lett., 2737 (1974).
291. T. Itoh, K. Kaneda and S. Teranishi, Bull. Chem. Soc. Japan, 48, 1337 (1975).
292. R. Sheldon and J. Van Doorn, J. Organometal. Chem., 94, 115 (1975).

FUNCTIONALIZATION OF OLEFINS

Guido P. Pez and Samir T. Bustany

Allied Chemical Corporation

Morristown, New Jersey 07960

Our task in this workshop is to see how developments in homogeneous catalysis might be applied towards the solution of energy-related problems in the chemical industry. Energy savings could be realized by the use of less costly feedstocks and by the employment of more direct, simple and efficient chemical processes. This talk reviews some recent developments in the catalytic chemistry of olefins and suggests desirable novel chemistry that would, if realized, bring inportant saving in energy or feedstocks. In order to put this discussion in a realistic economic perspective, we will first discuss the present and future sources, availability, and economics, of the three major olefins: ethylene, propylene and butadiene.

Feedstocks and Economics

Before World War II, acetylene from coal was the foundation of the chemical industry, particularily in Europe. Since then, olefins from oil or natural gas have progressively replaced acetylene and become the primary building blocks of the chemical industry. Table I shows that in 1976 there were 23.5 billion pounds of ethylene produced in the U.S. and only one half billion pounds acetylene. Acetylene today is more than twice as costly as the two preeminent olefins, ethylene and propylene, principally because of a much lower yield in its synthesis from natural gas. Saturated hydrocarbons are cheaper than olefins; unfortunately, we do not as yet know how to functionalize them efficiently or directly, so we crack them to olefins which then are functionalized. Finally, there is a lot of recent interest in functionalizing CO/H_2 mixtures, i.e., synthesis gas. The reason is that CO/H_2 is made fron steam re-

Table 1. Major building blocks of the chemical industry.

Feedstock	U.S. Production (1976) (10^9 lbs/year)	Price (1976) (¢/lb.)
HC≡CH	0.5	25-30
$H_2C≡CH_2$	23.5	12
$H_2C=CH-CH_3$	9.5	8.5
$H_2C=CH-CH=CH_2$	3.5	17
CO/H_2	38	3?

forming of natural gas or naphtha and is mostly used for making
ammonia and methanol. In the 21st century a substantial fraction
will be made from coal. In the transition period the economics
of CO/H_2 are uncertain. In the most favorable case, CO/H_2 will be
supplied as a slipstream of a huge coal gasification plant with
utility-type financing and government subsidies.

The volume of olefin demand depends on their price and availability.[1] Our forecast for olefin prices is shown in Figure I.

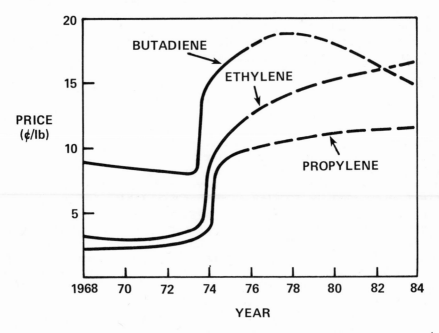

Figure I. Price forecast of Olefins (in constant 1976 dollars).

The writers do not want to indulge in futurology, but rather to provide the reader with a useful framework. Ethylene price dropped in the 1960's to a low of 3¢/lb. because of larger plants and over capacity. Traditionally in the U.S., ethylene has been produced by cracking ethane, which is produced from natural gas liquefaction. Due to the expected short supply of natural gas, the majority of the new ethylene plants will be based on cracking of heavier feedstocks, namely naphtha and gas oil. The first consequence of this is that ethylene will not be the very cheap building block that it used to be. A second important consequence is the large amount of propylene, butadiene and other C_4 coproducts generated in these new ethylene plants. This additional supply of propylene should cause a spread spmewhat larger than today's between the price of ethylene and propylene. This, in turn, will result in a fast growth of propylene derivatives. Butadiene will go down in price because of the over-supply of coproduct butadiene. The extent of the butadiene price decline will depend on whether new uses for this olefin are found.

The present major uses for olefins are shown in Table 2. It is clear that polymers including fibers represent the largest applications of olefins: polyethylene, polyester fibers, polyvinyl chloride from ethylene; polypropylene and acrylic fibers from propylene; synthetic rubber and hexamethylene diamine, (a precursor for Nylon-6,6) from butadiene.

Table 2. Major olefin derivatives.

Olefin		1974 U.S. Production (in 10^6 lbs/year)
Ethylene	Polyethylene	8,800
	Vinyl Chloride	5,700
	Ethylene Oxide/Glycol	3,900
Propylene	Polypropylene	2,200
	Acrylonitrile	1,400
	Butyraldehyde	600
Butadiene	Rubber, Polymers	3,100
	Adiponitrile/HMDA	400*

* About another 400 MM lbs/year not from butadiene.

Catalytic Processes

Olefins are basically useful because of their wide scope of catylytic chemistry. The major industrial olefin derivatives have already been cited. Our aim in this section is not to review this technology in detail, but rather, to focus on its basic chemistry and see how process improvements, or basically new olefin technology might be realized, in the future.

Aside from polymerization, most olefin reactions involve the addition of heteroatomic species to the unsaturated moiety. The catalytic chemistry with olefins can be divided into two broad categories: These are,

A. Direct heteroatom additions, and

B. Oxidative olefin additions

An example of the former is the simple hydration of an olefin to an alcohol. We envisage reactions of the second type as those which involve oxygen, whether or not an oxygen funstion is incorporated into the final product. The well-known Wacker process for the conversion of ethylene and oxygen to acetaldehyde, and the ammoxidation of propylene and ammonia and air to acrylonitrile, are illustrations of this type of reaction. One reason for making this classification is that in practice, the catalytic chemistry of the two types is very different. Type A processes involve, essentially, reductive chemistry, and are done in the presence of excess reducing reagents (e.g., CO, H) with Group 8 metals, or, with low-valent (and often air sensitive) metal complex catalysts. In contrast, the oxidative type B transformations are mostly carried out over early transition metal oxides (e.g. V_2O_5) as well as over the later transition (more noble) metals themselves (e.g. Pt, Ag). More recently developed, homogenous olefin oxidation catalysts range from Group 8 complexes (mostly Rh), to non-metal redox systems in aqueous or strong acid media. Attention will now be focused on a discussion of specific catalytic systems.

A. Direct Heteroatom Additions

Reaction systems to be discussed are the following:
(I) "Bare Metal Atom" Catalyst Systems.
(II) Hydride, Alkyl and Carbonyl-Metal Catalysts.
(III) Ammination of Olefins

I. "Bare Metal Atom" Catalyst Systems

The simplest system which can be envisioned is that of a metal atom, that is stabilized by ligands that are themselves reagents

(or products) of a catalytic reaction. A now classical example, is
the cyclooligomerization of three molecules of 1,3-butadiene around
nickel to yield mainly the trans, trans, trans cyclododecatriene
(CDT)[2]

$$3 \diagup\!\!= \quad + \quad \xrightarrow[\text{(Wilke)}]{\text{"Naked Ni"}} \quad \xrightarrow{C_4H_6}$$

The synthesis proceeds via the well characterized bis $(\eta\text{-allyl})\text{-}C_{12}$
nickel intermediate which upon further reaction with 1,3-butadiene
yields free CDT. The various so called "naked nickel" catalysts
(Wilke) are prepared by the reduction of a nickel salt in the presence of diene; a standard catalyst combination is nickel acetylacetonate with diethylaluminum ethoxide. Atomic nickel prepared
by metal evapor ation techniques is also effective. Such "bare
atom" catalysts as are active for the trimerization of 1,3-dienes
can be modified by the addition of one ligand, to now function
as 1,3-diene dimerization catalysts. Thus, reaction of a 1:1 Ni:
triphenylphosphite ("nickel- ligand catalyst (Wilke) at 80°C
yields cyclooctadiene together with smaller quanties of 4-vinyl-
cyclohexene and CDT.[2]

With "palladium-ligand" systems 1,3-butadiene can be dimerized
and subsequently functionalized to yield octadiene derivatives of
alcohols, amines, and acids[3]:

$$2\ CH_2\text{=}CH\text{-}CH\text{=}CH \ + \ Y\text{-}H \xrightarrow{\text{Cat.}} Y\text{-}CH_2\text{-}CH\text{=}CH\text{-}(CH_2)_3\text{-}CH\text{=}CH_2 \ +$$
$$+ \ CH_2\text{=}CH\text{-}CH\text{-}Y\text{-}(CH_2)_3\text{-}CH\text{=}CH_2$$

This 1,3-butadiene telomerization reaction is thought to occur
by the external attack of a nucleophile Y-H on a bis $(\eta\text{-allyl})\text{-}C_8$
species, bound onto either one or two metal atoms. For the latter
case, the following mechanism has been proposed[3]:

$$4 C_4H_6 \ + \ HOAc$$
$$+ \ (\eta\text{-}C_3H_5)_2Pd$$

$$AcOCH_2\text{-}CH\text{=}CH\text{-}(CH_2)_3\text{-}CH\text{=}CH_2$$

Most known cyclization reactions with Group 8 "bare metal atom" or "metal-ligand" catalysts proceed with 1,3-dienes; very few cyclization reactions are known with ethylene. Ethylene can be trimerized to cyclohexane and mixtures of ethylene and 1,3-butadiene have been reported to yield vinylcyclobutane over some Ziegler-Natta catalysts.[4] However, such simple cyclization reactions as ethylene to cyclobutane[5] or ethylene and 1,3-butadiene to yield cyclohexadiene, have (to the best of our knowledge) not yet been realized. The cyclization of 1,5-hexadiene to cyclohexane, would be another challenge for new "bare metal atom" catalysts.

The catalyst systems that we have treated thus far are based on Group 8 metals; well-characterized "bare atom" catalysts of the earlier transition metals are rare, and far more difficult to prepare. The reason is that these metals have relatively empty d-shells so that a full complement of good donor ligands is usually required for the isolation of stable complexes. Despite this, by using a novel preparative technique, we have been able to isolate a highly metal unsaturated complex catalyst: μ-(η^1: η^5-cyclopentadienyl)-tris(μ-cyclopentadienyl)dititanium (Ti-Ti), μ-(η^1: η^5-C_5H_4) (η-C_5H_5)$_3$Ti$_2$:[6]

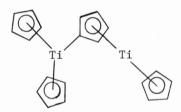

This "titanocene" compound has been characterized from a combination of its chemical properties and an x-ray structure of a bis-tetrahydrofuran derivative. The molecule is stable despite the fact that it possesses considerable electronic unsaturation (EAN=29). A striking feature of the molecule is the presence of a considerable degree of openness or structural unsaturation caused by a "bending back" of the Ti-Ti linkage by the bridging (η^1: η^5-C_5H_4) ligand.

It may seem unusual that a molecule with such a degree of saturation can actually be prepared. Our synthetic technique is to reduce titanocene dichloride $(C_5H_5)_2TiCl_2$ with potassium naphthalene at low temperatures:

$$(C_5H_5)_2TiCl_2 + 2K\ C_{10}H_8 \xrightarrow[-80°C]{THF} (C_5H_4)(C_5H_5)_3Ti_2$$

$$(40-60\%\ \text{yield})$$

In contrast, reduction of $(C_5H_5)_2TiCl_2$ at room temperature gives mainly the fulvalene bridging hydride: $\mu(\eta^5: \eta^5 - C_{10}H_8) - \mu H_2 - (C_5H_5)_2 Ti_2$. This ow temperature aprotic reduction technique has also been used to make zirconocene naphthalene, $[(C_5H_5)_2 Zr - C_{10}H_8]_n$ and we believe it to be a general technique for preparing "bare metal atom" catalysts.

The "titanocene" compound is an excellent catalyst for the isomerization and hydrogeneration of olefins. It is the only well-characterized complex of titanium to show catalytic activity with' olefins. However, isomerization and hydrogeneration reactions are well-known to occur with Group 8 metal complexes, and we felt that "titanocene" should display catalytic chemistry which would reflect its bimetallic highly unsaturated character. We have found that $(C_5H_4)(C_5H_5)_3Ti_2$ catalyzes the conversion of ethylene into ethane and 1,3-butadiene:

$$3\ C_2H_4 \xrightarrow[\text{THF}]{23°C} CH_2=CH-CH=CH_2 + C_2H_6$$

The reaction is unprecedented with either soluble or hetero-geneous catalysts. It differs from reactions of ethylene that have been observed over olefin metathesis catalysts. Unfortunately, the reaction is very slow but it is definitely catalytic. The synthesis of 1,3-butadiene from ethylene probably proceeds via the formation of a dititanium metallocycle, on the "open" side of the Ti-Ti linkage of our "titanocene" compound:

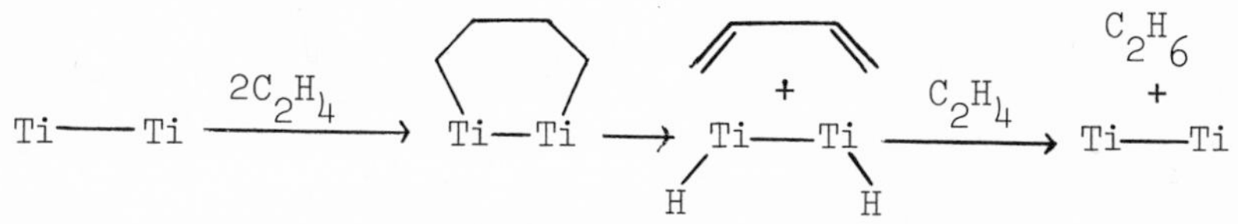

We feel that such novel (and often unpredictable) catalytic chemistry of olefins will be found over "bare metal atom" catalyst systems – especially those of the earlier transition metals.

II. Hydride, Alkyl and Carbonyl Catalysts

Most industrial catalysts employed for the hydrogenation, oligomerization, polymerization, and hydroformylation of olefins fall under this category. A basic transformation in all of these processes is the reversible synthesis of metal alkyls from olefins and metal hydrides. The reaction of a metal hydride with an olefin may be thought of as cccuring in two steps, namely (a) olefin coordination, and (b) hydride transfer. Olefin coordination is expected to occur best with low-valent (coordinatively unsaturated) metal hydrides. The conditions for the efficient synthesis of stable

$$\underset{M}{\overset{H}{|}} + \underset{CH_2}{\overset{CH_2}{||}} \rightleftharpoons \underset{M}{\overset{H}{\diagdown}}\diagup\!\!\!\!\diagup \rightleftharpoons M-CH_2CH_3$$

metal alkyls are not as clear. Metal alkyls are formed stoichiometrically from $(C_5H_5)_2ZrHCl$[9], and from trans $(PPh_3)_2PtHCl$[10]. With $(PPh_3)_3RuHCl$ and $(PPh_3)_3RuHOAc$, the corresponding alkyl species eg. $(PPh_3)_3Ru(C_2H_5)Cl$[11] are found in small equilibrium amounts, under pressure of ethylene. In our experience, discrete, low-valent titanium hydrides yield mainly oligomeric alkyls on reaction with ethylene. Much needs to be done, before we can better appreciate the process of hydride transfer onto coordinated olefins, as well as the reverse process of β–hydrogen elimination.

Perhaps the most desirable fundamental study, in this regard, would be to measure the Bronsted acidity of transition metal hydrides. Hydrides of titanium and zirconium, for example, $(C_{10}H_8)(C_5H_5)_2Ti_2H_2$[12] and $|C_5(CH_3)_5|ZrH_2$ are usually considered to be "hydridic"[5]. (The latter[5] has recently been shown to be capable of reducing CO to $CN_2OH)$[13]. In contrast, hydrido cobalt tetracarbonyl and related Fe, Mn, V species behave as acids in aqueous media.[14] There are also many other hydrides, as for example, $(PPh_3)_4RuH_2$ and $(PPh_3)_3RhH_2Cl$ which cannot at present be readily classified as displaying either acidic or hydridic behavior. It would certainly be instructive, for the purpose of developing novel olefin-reactions catalysts, to quantify the Bronsted acidity of a range of transition metal hydrides in a series of common solvent systems.

Metal alkyls, produced from olefins and hydrogen, are readily functionalized by carbon monoxide: this is the familiar hydroformylation reaction for the synthesis of aldehydes and alcohols. However, direct ethylene telomerization processes, i.e.,

$$nC_2H_4 + \psi-H \longrightarrow CH_3-(CH_2)_{2n-2}-CH_2-\psi$$

(where $\psi-H$ = alcohols, amines etc.)

are not known. Such processes might eventually be realizable from a better understanding of the chemistry of transition metal hydride, aklyl, and carbonyl complexes.

Hydroformylation chemistry depends essentially on insertion of carbon monoxide into an olefin derived metal-aklyl linkage. Recent work with carbonyl osium clusters, suggests that another kind of CO-insertion might be possible. Reaction of $Os_3(CO)_{12}$ with hydrogen gives the hydrido-cluster $H_2Os_3(CO)_{10}$.[15] Treatment with ethylene yields a hydrovinyl complex and ethane:

$$H_2Os_3(CO)_{10} + 2C_2H_4 \xrightarrow{25°} (CO)_3 \underset{\text{Os}}{\triangle}(CO)_3 + C_2H_6$$

The hydrovinyl complex of formula $HOs_3(CO)_{10}CH=CH_2$ may be considered as arising from a formal insertion of osium into the C-H bond of ethylene: This novel chemistry suggests that with some carbonyl clusters it might be possible to catalyze reactions involving the insertion of CO into olefinic C-H linkages. An example, would be the direct carbonylation of ethylene to acrolein:

$$CH_2=CH_2 + CO \longrightarrow CH_2=C \begin{smallmatrix} H \\ \\ CHO \end{smallmatrix}$$

An obvious problem to be overcome is that with such a system it may be difficult to generate a sufficient eduilibrium concentration of the unsaturated metal cluster, in the presence of reagent carbon monoxide.

III. Ammination of Olefins

The homogenous hydration of ethylene to ethanol proceeds easily with acidic catalysts, yet the direct ammination of ethylene with ammonia:

$$NH_3 + C_2H_4 \longrightarrow C_2H_5NH_2 \qquad K_{25°C} \sim 230$$

though thermodynamically feasible[17] has barely been realized. Ethylene and ammonia are reported[16] to react in the presence of sodium metal under extreme conditions (400 atm, 200°C) to yield ethylamine. Ethylamine is also obtained by heating C_2H_4 and NH_3 over a supported molybdenum oxide at ~200°C. However, at this temperature the reaction is slow, while at higher temperatures the equilibrium becomes unfavorable. No homogeneous catalyst for the addition of NH_3 to ethylene or to alpha-olefins has yet been developed. The stoichio-metric addition of NH onto C_2H_4 that has been "acidified"[18] by co-ordination onto Pt+2, has however, been demonstrated:

$$C_2H_4-PtCl_2 + NH_3 \longrightarrow [H_3\overset{\oplus}{N}-CH_2-CH_2-\overset{\ominus}{PtCl_2}]$$

$$\xrightarrow{NH_3} [H_2N-CH_2CH_2-PtCl_2]^{\ominus} + NH_4^{\oplus} \xrightarrow{H^+} C_2H_5NH_3^{\oplus}Cl^{\ominus}+NH_4Cl$$

The addition of secondary amines to ethylene and to 1,3-butadiene is also known:

$$C_2H_4 + R_2NH \xrightarrow[\text{or Li(TMED)N}(C_2H_5)_2]{\text{Rh cat.}} R_2N(C_2H_5) \quad \text{Ref. 19,20}$$

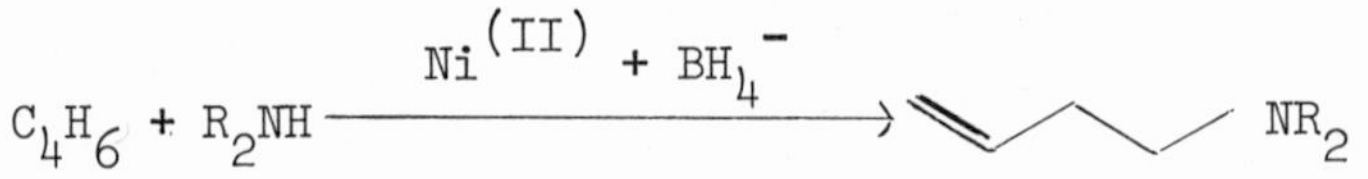

$$C_4H_6 + R_2NH \xrightarrow{Ni^{(II)} + BH_4^-} \diagdown\diagdown\diagup\diagup NR_2 \quad \text{Ref. 21}$$

These reactions probably proceed via a nucleophilic addition of an R_2N^- moiety onto the olefin (this must certainly be the case with the Li (TMED) $\bar{N}(C_2H_5)_2$ catalyst).[20] Two possible mechanisms (and their problems) are envisioned for achieving a catalytic addition of NH_3 onto olefins:

(a) Addition of NH_2^- from KNH_2 onto a coordinated olefin:

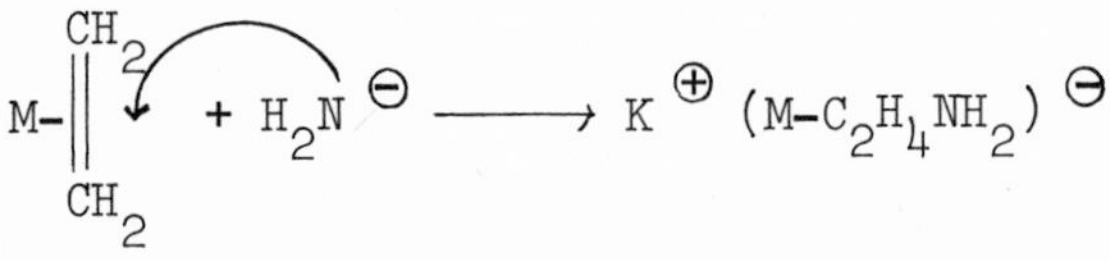

$$M-\underset{CH_2}{\overset{CH_2}{\|}} + H_2N^{\ominus} \longrightarrow K^{\oplus} (M-C_2H_4NH_2)^{\ominus}$$

followed by protonolysis of the metal alkylamide anion to yield the metal-olefin complex, potassium amide and the product alkylamine:

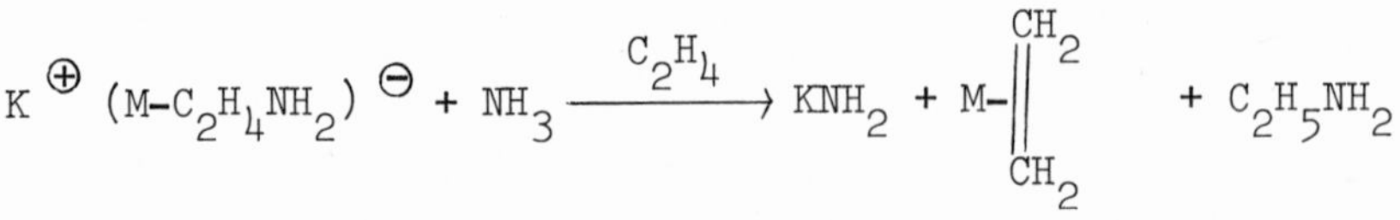

$$K^{\oplus} (M-C_2H_4NH_2)^{\ominus} + NH_3 \xrightarrow{C_2H_4} KNH_2 + M-\underset{CH_2}{\overset{CH_2}{\|}} + C_2H_5NH_2$$

A difficulty with this potential M/KNH_2 catalyst system is that very few transition metal complex olefin catalysts retain their integrity in the presence of as strong a base as potassium amide.

(b) Amide insertion into a metal-alkyl bond:

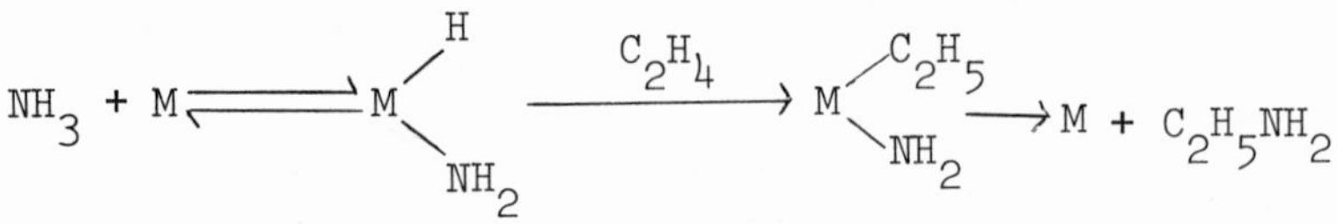

$$NH_3 + M \rightleftharpoons M\overset{H}{\underset{NH_2}{\diagup}} \xrightarrow{C_2H_4} M\overset{C_2H_5}{\underset{NH_2}{\diagup}} \longrightarrow M + C_2H_5NH_2$$

The problem here is that oxidative addition reactions of NH_3 (and aliphatic amines) have not yet been documented. However, J. Armor in these laboratories[22] has shown that "titanocene" $(C_5H_4)(C_5H_5)_3Ti_2$

will spontaneously react with NH_3 to produce a hydridobis(imide) bis(dicyclopentadienyltitanium) complex and hydrogen:

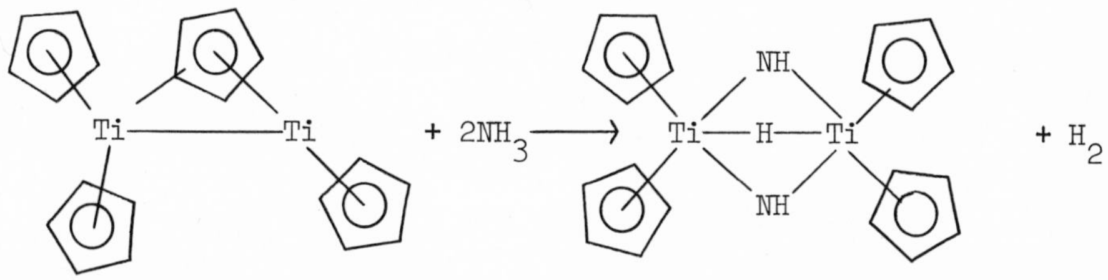

It is thought that the formation of this imide complex and the accompanying evolution of hydrogen arises from an oxidative addition of NH_3 onto the metal centers of the "titanocene". (A similar reaction of NH_3 with clean iron surfaces to yield surface amides and imides and H^3-gas has been noted).[23] Reaction of this "titanocene" with ammonia and ethylene yields only stoichiometric quantities of ethylamine. the required $M\diagdown{}^{H}_{NH_2}$ intermediate postulated earlier, is probably formed in this "titanocene" system. However, the metal hydride, instead of reacting with ethylene, reacts with a second mole of NH_3 to yield the dititanium diimide and hydrogen, so that a catalytic NH_3/C_2H_4 reaction is not achieved.[24]

B. Oxidative Heteroatom Additions

Oxidative olefin functionalization reactions are essentially of two types: Olefins may be oxidized directly to epoxides, aldehydes and ketones; of alternatively oxygen can be employed indirectly in a redox system to partially oxidize a reagent for functionalization of the olefin. Briefly, technology for the direct partial oxidation of olefins includes the conversion of ethylene to ethylene oxide, over silver-surface catalysts; the oxidation of propylene or propylene/propane mixtures to propylene oxide, and the homogeneous Wacker process for the oxidation of ethylene to acetaldehyde. Most of this technology is now well-developed; remaining problems associated with it relate to improving the selectivity (for the epoxides) and on minmizing reaction corrosion (for the Wacker process). Recently, homogeneous, metal complex catalyzed, non-radical oxidations have been reported. An example is conversion of alpha-olefins to aldehydes and ketones in the presence of tris(triphenlphosphine)rhodiumchloride:[25]

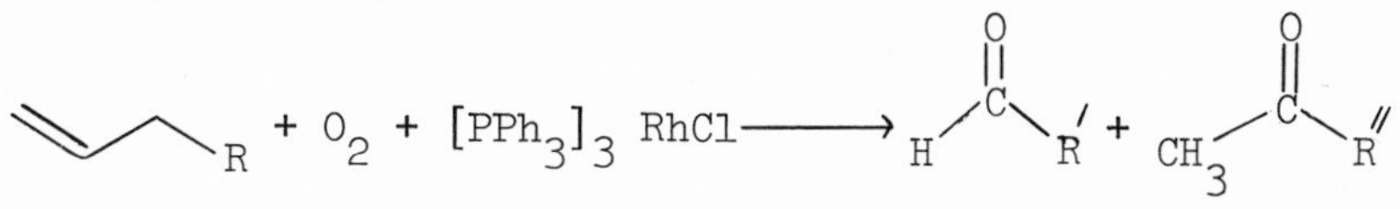

An obvious difficulty with metal-complex induced oxidations is that the ligand (often a phophine) is inevitably itself oxidized. This might be avoided, and in time, we could see the development of highly selective homogeneous oxidation catalysts. A comprehensive review of partial oxidation chemistry is given elsewhere in this volume by J. Lyons.

The most important indirect oxidative heteroatom addition re-actions are the ammoxidation of propylene to acrylonitrile:

$$CH_3CH=CH_2 + NH_3 + O_2 + \frac{Bismuth}{Molybdate\ Cat.} \longrightarrow CH_2 = CH-CN;$$

and the recently developed homogeneous oxidation hydration of ethylene to ethylene glycol:

$$CH_2=CH_2 + 1/2\ O_2 + H_2O \longrightarrow HOCH_2CH_2OH$$

We will describe this system in some detail since it may form the basis of other novel olefin functionalization processes. At the present time, ethylene glycol is made in two steps. In the first step, ethylene is catalytically oxidized with air to yield ethylene oxide and carbon dioxide. The ethylene oxide id hydrated in the second step. The principle disadvantage of this process is the relatively low yield (65-75%) due to CO_2 formation. This has pro-moted many companies to look for routes with a better yield. Most of these processes comprise acetoxylation of ethylene to mono and diacetates followed by hyrolysis to yield over 90%. This company, jointly with Arco is building a 800MM lbs ethylene glycol plant in Channelview, Texas, based on this process. The catalyst is tel-lurium oxide and aqueous HBr and 2-bromoethylacetate. A probable reaction mechanism is the following:

$$CH_2=CH_2 + Br_2 \longrightarrow BrCH_2CH_2Br$$

$$BrCH_2CH_2Br + 2HOAc \longrightarrow AcOCH_2CH_2OAc + 2HBr$$

$$Te^{+6} + 2HBr \longleftarrow Te^{+4} + Br_2 + 2H^+$$

$$2H^+ + 1/2\ O_2 + Te^{+4} \longrightarrow Te^{+6} + H_2O$$

$$CH_2=CH_2 + 1/2\ O_2 + 2HOAc \longrightarrow AcOCH_2CH_2OAc + H_2O$$

Kuraray's process is similar to Halcon's, but uses a palladium instead of a tellurium catalyst. Nitric oxides are used to reoxidize reduced Pd^0 to Pd^{+2}. The principal product is the monoacetate. Teijin, another Japanese company, has developed the following one-step process where ethylene chlorohydrin is an intermediate:

$$CH_2=CH_2 + TlCl_3 + H_2O \longrightarrow ClCh_2CH_2OH + TlCl + HCl$$

$$ClCH_2CH_2OH + H_2O \longrightarrow HOCH_2CH_2OH + HCl$$

$$TlCl + 2CuCl_2 \longrightarrow TlCl_3 + Cu_2Cl_2$$

$$2\ Cu_2Cl_2 + 4HCl + O_2 \longrightarrow 4\ CuCl_2 + 2H_2O$$

Recent patents[27] even describe the highly selective hydration of 1,3-butadiene to 1,4-butanediol, using a $PdTe_4$ cluster catalyst.

It is conceivable that other redox systems might be found that will permit the oxidative ammination of ethylene to ethylene diamine:

$$CH_2=CH_2 + 2NH_3 \xrightarrow[\text{Cat.}]{O_2} H_2NCH_2CH_2NH_2 + H_2O$$

This unknown reaction would be very attractive since the present ethylene diamine process involves two steps, a yield of only 70% and the degradation of chlorine to HCl:

$$CH_2=CH_2 + Cl_2 \longrightarrow ClCH_2CH_2Cl$$

$$ClCH_2CH_2Cl + 2NH_3 \longrightarrow NH_2CH_2CH_2NH_2 + 2HCl$$

There is some precedent for an oxidative addition of ammonia to olefins; as in the following (non-catalytic)[28] conversion of ethylene and ammonia to α-picoline:

$$C_2H_4 + NH_3 + Pd(NH_3)_4Cl_2 + H_2O \xrightarrow{120°C} \text{(α-picoline)} + PD^{(o)}?$$

In general, it may be expected that more competitive, novel routes to organo-nitrogen, silicon, phosphorous, and sulphur chemicals will be developed by the catalytic oxidative addition of NH_3, silane, phosphine, hydrogen sulphide, etc. to olefins.

Conclusions

At present, half or more of the organic chemicals are produced by the functionalization of olefins. Certain recent papers might lead one to believe that the "olefin age" is over. We believe this is not so. Acetylene is too costly, alkanes are too difficult to functionalize, and CO/H_2 has its own uncertainties. Olefins will remain the most important building blocks of hte chemicla industry for the next 20 years. What is true is that ethylene will not be as cheap as it was in the 1960's. Olefin costs will represent a major, if not the major component of the total costs. Therefore, more research will be directed towards processes with very high yield on olefin.

We have discussed the catalytic chemistry of olefins in terms of (a) direct, and (b) oxidative heteroatom addition reactions. Hydrolysis and hydroformylation are among the most important direct olefin functionalization processes. On the other hand, such direct additions as, for example, NH_3 or CO the ethylene (to yield ehtylamine and acrolein, respectively) are presently unknown, and represent a challenge for catalysis research. The so-called "bare atom", metal hydride and alkyl and metal carbonyl catalysts types have been discussed. We feel that such novel (and often unpredictable) chemistry of olefins will result from research on "bare metal atom" systems, especially thos of the earlier transition metals.

Several examples of oxidative heteroatom addition reactions have been described, and in particular the Halcon route to ethylene glycol. As on extension of this highly successful process, we envisage that organonitrogen, silicon, phosphorous and sulphur compound will, in the future, be produced by oxidative addition of NH_3, silane, phosphine, etc., to olefins.

References

1. U.S. International Trade Commission, "Synthetic Organic Chemicals" Washington, D.C. 20436.
2. P.W. Jolly and G. Wilke, "The Organic Chemistry of Nickel", Vol. II, p. 133, Academic Press (1975).
3. W. Keim, "π-Allyl System in Catalysis" in "Transition Metals in Homogeneous Catalysis", p. 68, G. Schrauzer, Ed., Marcel Dekker, Inc. (1971).
4. L. G. Cannell, J. Amer. Chem. Soc., 94, 6867 (1972).
5. F. D. Mango and J. H. Schachtschneider in "Transition Metals in Homogemeous Catalysis", p. 225, G. Schruazer, Ed., Marcel Dekker, Inc. (1971).
6. G. P. Pez, J. Amer. Chem. Soc., 98, 8072 (1976).
7. G. P. Pez and S. C. Kwan, J. Amer. Chem. Soc., 98, 8079 (1976).

8.   G. P. Pez, J.C.S. Chem. Comm., submitted for publication (1977).

9.   J Schwartz and J. A. Labinger, Angewandte Chemie, International
     Ed., 15, 333 (1976).

10. J. Chatt and B. L. Shaw, J. Chem. Soc., 5075 (1962).

11. P. S. Hallman, B. R. McGarvey and G. Wilkinson, J. Chem. Soc.,
     A, 3143 (1968).

12. A. Davison and S. S. Wreford, J. Amer. Chem. Soc., 96, 3017
     (1974), and references therein.

13. J. M. Manriquez, D. R. McAlister, R. D. Sanner and J. E. Bercaw,
     J. Amer. Chem. Soc., 98, 6733 (1976).

14. R. A. Schunn, "Systematics of Tranition Metal Hydride Chemistry"
     in "Transition Metal Hydrides", p. 239, E. L. Muetterties,Ed.,
     Marcel Dekker, Inc. (1971).

15. (a) J. B. Keister and J. R. Shapley, J. Amer. Chem. Soc., 98,
            C29 (1975), and references therein.
     (b) J. B. Keister and J. R. Shapley, J. Amer. Chem. Soc., 98,
     1056 (1976).

16. B. W. Howk, E. L. Little, S. L. Scott and G. M. Whitman, J.
     Amer. Chem. Soc., 76, 1899 (1954).

17. L. Uhlmann, Wissenschaftliche Zeitschrift der Technischen
     Hochschule Für Chemie Leuna-Merseburg, 5, 263 (1963).

18. A. De Renzi, G. Paiaro, A. Panunzi and L. Paolillo, Gazzetta
     Chimica Italiana, 102, 281 (1972).

19. D. R. Coulson, Tet. Lett., No. 5, 429 (1971).

20. H. Lehmkuhl and D. Reinehr, J. Organomet. Chem., 55, 215 (1973).

21. R. Baker, D. E. Halliday and T. N. Smith, J.C.S. Chem. Comm.,
     1583 (1971).

22. J. Armor, Submitted to J. Amer. Chem. Soc., (1977).

23. D. O. Hayward and B. M. W. Trapnell, "Chemisorption", p. 247,
     Butterworths (1964).

24. J. Armor, private communication (1977).

25. C. W. Dudley, G. Read and P. J. C. Walker, J. Chem. Soc.,
     (Dalton) 1926 (1974).

26. Chem Systems, Inc., Report No 74-1 "Ethylene Glycol" (1974).

27. U.S. Patent No. 3,922,300 (1975).
     Belgium Patent No. 832,254 (1976).

28. Japanese Patent No. 71 09,586 (1971) CA 75, 20213y (1971).

CATALYSIS BY SUPPORTED COMPLEXES

J. M. Basset and A. K. Smith

Institut de Recherches sur la Catalyse

79, bd du 11 novembre 1918 69626-Villeurbanne-France

Since about 1970 there has been a growing interest in the literature concerning catalysis by supported complexes. The starting goal was the following: when one grafts an homogeneous catalyst to a support it should be possible to accumulate the advantages of homogeneous catalysis, namely activity and selectivity and the advantages of heterogeneous catalysis, namely the easy separation and re-use of the costly catalyst. Besides these starting goals which have been reached in many cases, the subject of catalysis by supported complexes which overlaps homogeneous and heterogeneous catalysis, has been at the origin of new ideas, new synthesis, new concepts and unexpected results in both fields, so that at the present time neither homogeneous nor heterogeneous catalysis can ignore the real possibilities of these new types of catalysts which are already in industry the third generation of catalysts.

The main purpose of this report is not to make a complete survey of the literature in this field, since many excellent reviews already appeared in the recent years.[1-10] Nevertheless we would like to show what are the trends now in this field with a particular emphasis on the most interesting results which are typical and can define by themselves new areas which might deserve new developments. It must be pointed out first that supported enzymes have been at the origin of this field and this is a subject of intensive research in biochemistry. Nevertheless we will not consider supported enzymes here, although it should be very useful in the future to keep a scientific overlap between both fields.

If one considers the field of catalysis by supported complexes in the last six years many trends have been developed which represent in fact the purpose that one wanted to reach.

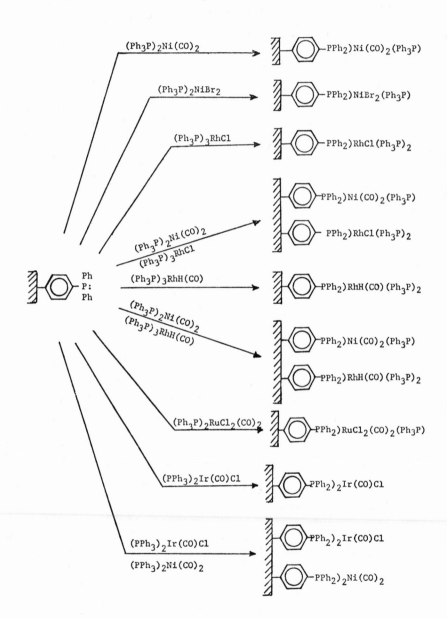

FIG. 1.

The first approach consisted to graft an homogeneous catalyst onto a support and try as a primary goal to keep its molecular nature hoping thus to preserve its activity and selectivity. In this respect two main approaches have been tried:
- grafting the complex on a polymer support soluble or not,
- grafting the complex on an inorganic oxide.

$$\boxed{support} \longrightarrow L \rightarrow ML_x$$

The second approach consisted to support an homogeneous catalyst onto a support and try as a primary goal to reach the activity and selectivity as high as possible without any care about the fact that the molecular nature of the catalyst is kept or not. In this case most of the works have dealt with inorganic oxides as supports and two main approaches have been tried:
- direct interaction of an homogeneous catalyst onto a support

$$\boxed{support} \sim\!\sim\!\sim ML_x$$

- reaction of an organometallic with the functional groups of a support with at least a partial loss of its molecular nature

I.   The Molecular Nature of the Catalyst is Kept

I.A.   The polymer as a support. This field has certainly been the most intensively studied from a fundamental point of view. A great deal of work has been carried out by Grubbs et al.[11-13] as well as by Pittmann et al.,[14-18] Braca et al.,[19] Graziani et al.[20] Using polystyrene-divinyl benzene cross linked polymers the most common technique used consists to link a phenyl phosphine group to a benzene ring of the polymer as indicated in Figure 1.[21] Then by simple ligand exchange it is possible to graft onto the polymer e.g. zerovalent nickel complexes, bivalent nickel complexes, the Wilkinson complex, ruthenium[II] complexes, $RhH(CO)(PPh_3)_3$, the VASKA complex and so on. Many catalytic reactions have been carried out with these supported complexes among them olefin hydrogenation has received the greatest attention. However Pittmann and co-workers[21] have also studied in details butadiene oligomerization leading to cyclic dimers and trimers, or to linear dimers (Fig. 2), with supported nickel.

An interesting feature of these supported complexes is related to the possibility of multifunctionality obtained by grafting on the same polymer two different catalysts able to carry out two dif-

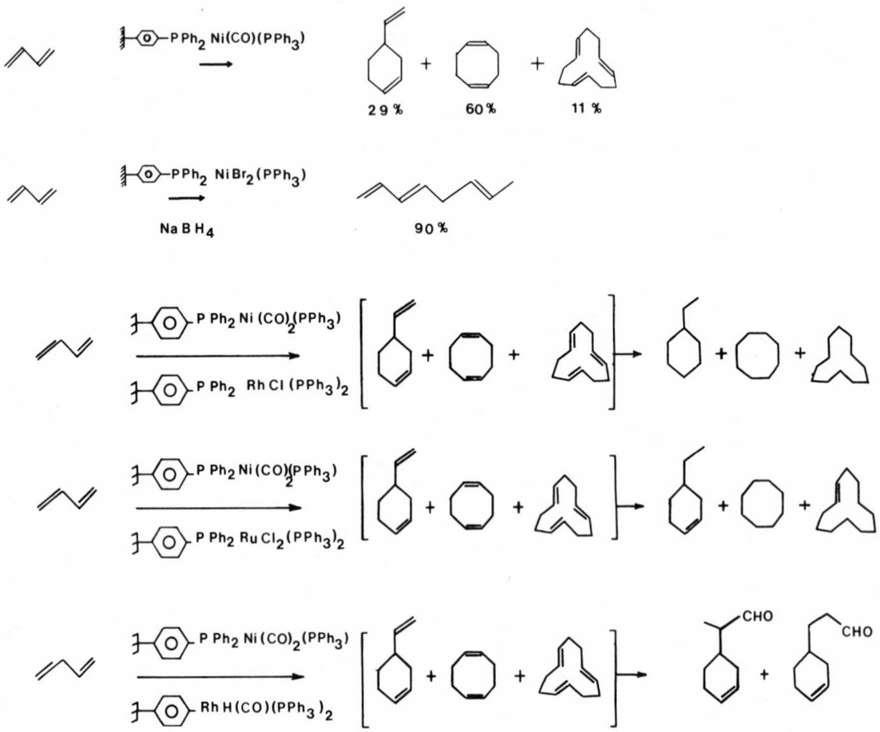

FIG. 2. Catalytic reactions with polymer anchored catalysts.

ferent reactions in a two step process. For example, it is possible
to cyclo oligomerize butadiene on a nickel[0] complex and then to
hydrogenate the cyclic dimers and trimers to the saturated hydrocarbons with the WILKINSON catalyst. In the same way the cyclic
dimers or trimers can be selectively hydrogenated with a grafted
Ru[II] complex. Finally vinyl-cyclohexene alone can be selectively
hydroformylated with a rhodium, carbonyl hydrido complex into linear
and branched aldehyde. This multifunctional process represents
certainly an interesting approach of catalysis by supported complex but it should be pointed out also that the performances obtained certainly do not overcome what should be expected with a
two stages process with two grafted catalysts.

An important contribution to the field of multifunctional catalysis by supported complexes has been brought by Gates et al.[22]
who applied to the "ALDOX" process a multifunctional catalyst which
consists of a rhodium[I] complex and amine groups linked to a polystyrene-divinyl benzene matrix. This industrially important process is a multistep synthesis involving homogeneous catalysis.
Propylene is hydroformylated to give butyraldehydes in the presence
of a Co or Rh catalyst, the aldehydes undergo base catalysed aldol
condensation and catalytic hydrogenation of the condensation products yields 2-ethyl hexanol, a major plasticizer alcohol:

$$CH_3-CH=CH_2 \xrightarrow[\substack{CO+ \\ H_2}]{Rh} \begin{array}{c} CH_3-\overset{\overset{\displaystyle CH_3}{|}}{CH}-CHO \\ \\ CH_3-CH_2-CH_2-CHO \end{array} \xrightarrow[\text{Amine}]{-H_2O} CH_3-\overset{\overset{\displaystyle CH_3}{|}}{CH}-CH=\underset{\underset{\displaystyle CH_3}{\underset{|}{CH_2}}}{\overset{|}{C}}-CHO$$

$$\Big\downarrow H_2 \quad Rh^I$$

$$CH_3-\overset{\overset{\displaystyle CH_3}{|}}{CH}-CH_2-\underset{\underset{\displaystyle CH_3}{\underset{|}{CH_2}}}{\overset{|}{CH}}-CHO$$

A representative structural element of the multifunctional
catalyst is given on Figure 3. The multifunctional catalyst beads
were compared to a combination of two kinds of catalyst beads, one
containing only Rh complexes and the other containing only amine
groups. The multifunctional catalyst was superior, giving rate
constants which were five-fold greater for hydroformylation, 15-
fold greater for condensation and 30-fold greater for hydrogenation.
The obtained result which is remarkable by itself, has a more general

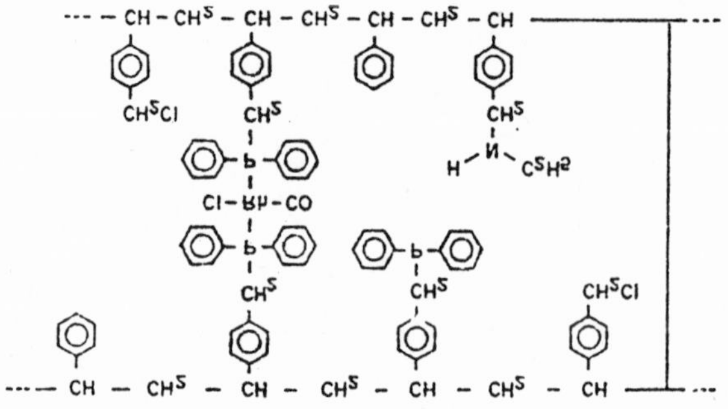

Figure 3.

interpretation which can be developed in the future; in a reaction
sequence such as the one written here:

$$(I) \quad \rightarrow \quad (II) \quad \rightarrow \quad (III) \quad \rightarrow \quad (IV)$$
$$\quad\quad X \quad\quad\quad\quad Y \quad\quad\quad\quad X$$

when both catalytic groups X and Y are bonded within a polymer matrix, the mass transfer resistance offered by the matrix can create
concentration gradients favourable to the second and third reactions
(condensation and hydrogenation) but unfavourable to the first (hydroformylation). In contrast when the sequence is catalyzed by
separately supported monofunctional catalysts, the concentration
gradients can only hinder all three reactions.

A more fundamental approach of these kinetics problems has
been given by many groups working with hydrogenation catalysts.
An important question which is by itself an oversimplification
could be the following: is there any change of the rate of an homogeneous reaction when the homogeneous complex is immobilized on a
polymer matrix? A partial answer was given by Grubbs[13] who studied
the reduction of cyclohexene with soluble and supported titanocene
associated with n-BuLi. The supported catalyst was found to be about
6 times more active than the homogeneous catalyst (Fig. 4). It
seems again that the result might have a more general implication
than the observed increased activity: grafting an homogeneous complex to a support may prevent bimolecular reactions leading to inactive dimers, or more generally inactive complexes. It might be a
good example which shows that immobilization of a complex on a support might stabilize coordinative unsaturation or unusual oxidation
states (e.g. Rh[II]).[20] Unusual rate effects have also been observed
by Pittmann,[17] in the hydrogenation of 1-5 cyclooctadiene catalyzed
by polymer anchored $IrCl(CO)(PPh_3)_2$. When compared at equal P/Ir
ratios the rate of hydrogenation was significantly faster using the
anchored catalyst whenever the P/Ir ratio was lower than 5. However
such results were not obtained by CHAUVIN et al.[23-24] in the case of
cyclopentene and acrylonitrile hydrogenation with a WILKINSON type
of catalyst supported on a soluble or an insoluble polymer matrix
(Fig. 5).

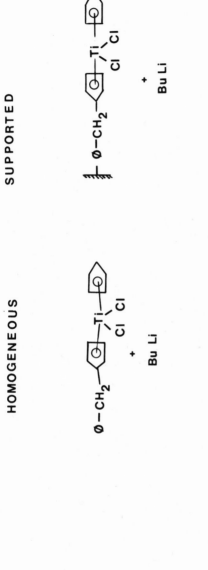

FIG. 4. Reduction of olefins with polymer bound and homogeneous titanocene.

Solvent	Catalyst	Initial rate mole H_2/min$\times 10^3$
C_6H_6	RhCl (P Ph$_2$ benz)$_3$	2.11
	supported soluble polystyrene	1.93
	supported insoluble $\left(\begin{array}{c}\text{20%cross-linked}\\\text{polystyrene-divinyl benzene}\end{array}\right)$	1.52
C_2H_5OH	RhCl (P Ph$_2$ benz)$_3$	5.33
	supported soluble polystyrene	3.97
	supported insoluble $\left(\begin{array}{c}\text{20%cross-linked}\\\text{polystyrene-divinyl benzene}\end{array}\right)$	1.98

FIG. 5. Comparison of activity supported/homogeneous cyclopentene hydrogenation.

With a soluble polystyrene, the rate of hydrogenation seems to be slightly lower than in pure homogeneous phase whatever the polarity of the solvent. When the polymer is highly cross-linked and insoluble the rate of hydrogenation is also much lower than in pure homogeneous phase. Obviously at the moment no general conclusions can be made concerning comparative rates of hydrogenation with homogeneous and supported complexes.

It appears that kinetics problems related to comparison between homogeneous catalysis and catalysis by supported complexes are very complicated and many parameters are playing a role simultaneously or separately.

Among these parameters diffusion of the substrate into the polymer "porous" system is certainly one of the most important. It appears now that this diffusion can be related to the size of the substrate as it occurs also in the framework of a zeolithe support. This might be purely a kinetic effect of diffusion. This size effect has been mainly studied by Grubbs.[12] Diffusion of the substrate into the polymer is also related to the chemical affinity of the substrate toward the support. This effect is clearly demonstrated in the catalytic hydrogenation of amino acid precursors (Fig. 6). Whereas the WILKINSON catalyst grafted on a polystyrene-divinyl benzene support does not result in the slightest activity for the phenyl-alanine hydrogenation, the presence of dimethyl amino groups on some phenyl ring of the polymer result in a good supported catalyst.

Another parameter which seems to be determining in catalysis by supported complexes is associated with the mobility of the polymeric chain which will determine the nature of the "average" coordination sphere of the grafted complex. Since there is a tremendous lack of structural information on what is really occuring in the polymer and since we have only a statistical estimate of the real nature of the grafted complex, any conclusion must be considered cautiously with respect to the coordination sphere of a grafted complex. Nevertheless works by Grubbs[13] and Collman[26] seem to indicate (Fig. 7) that the coordination sphere of $[Rh(COD)Cl]_2$ supported on a phosphinated polymer depends on the rigidity of the polymer, the mobility of the phenyl-phosphine groups being required to achieve the cleavage of the μ-chloro bridges. Nevertheless it is difficult to assess whether or not mobility exhibits a positive effect since rigidity is claimed to stabilize coordinative unsaturation. Besides, it seems that high mobility of the ligands can favor the multiple coordination of the ligands to the same metal center, increasing thus the extent of cross-linking in an unusual way, and favoring thus the insolubility of the polymer. Here again one does not know how mobile is a ligand in a polymer matrix differently cross linked. New

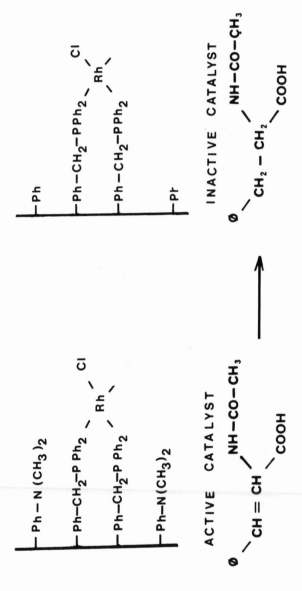

FIG. 6. Role of environment on catalytic activity of supported complexes.

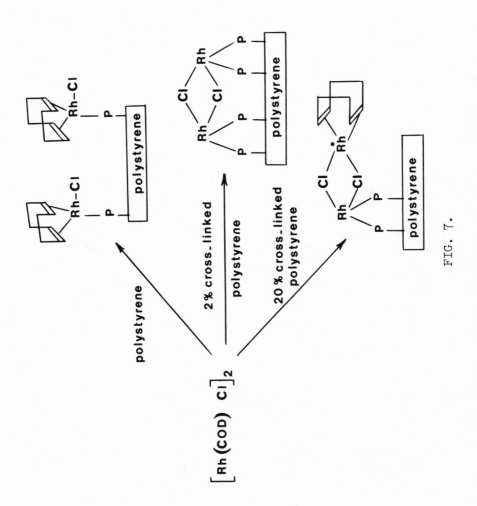

FIG. 7.

data have appeared in this field using the spin labeling technique[27] with paramagnetic nitric oxide covalently bonded to a polystyrene-divinyl benzene matrix. It has been shown that the degree of swelling of the polymer resin, as determined by the swelling solvent and cross-link density, has a substantial influence on the mobility of a nitroxide covalently attached to the resin. Besides it has been shown that the degree of swelling of cross-linked polystyrene beads in benzene has a significant influence on the rotational motion of a nitroxide imbibed in the solvent channels of the polymer, the magnitude of this effect is nearly equivalent to that observed for the same molecule incorporated into the polymer back-bone:

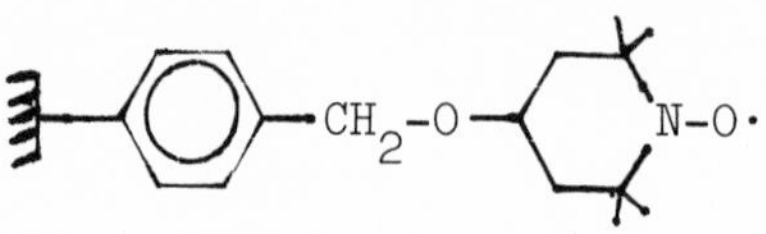

Finally the most interesting results obtained with polymer supported complexes are related to asymmetric catalysis by a supported chiral rhodium complex. The first result in this field was obtained by Kagan et al.[28] who succeeded in grafting the (DIOP) Rh^I catalyst on a Merrifield resin. This insoluble system catalyzes the asymmetric hydrogenation of a α ethyl-styrene and methyl atropate with a much lower efficiency than in solution. In contrast this supported catalyst is very efficient for asymmetric hydrosilylation of ketones (acetophenone, methylbenzylketone, isobutyrophenone). In this particular case the optical yield obtained with supported complexes are quite comparable to those obtained with an homogeneous (DIOP) Rh^I catalyst (Fig. 8), although the optical yield is usually lower with supported complexes.

In a more recent work Stille et al.[29] were able to support the (DIOP) Rh^I catalyst on a polymer which swells in polar solvents. The copolymer is obtained by free radical copolymerization of hydroxy ethyl metacrylate with a styrene unit having the (DIOP) ligand grafted in para position (Fig. 9).

The Rh^I catalyst grafted to this hydrophilic polymer is able to hydrogenate acyl amido acrylic acid in a polar solvent such as alcohol. The results of asymmetric hydrogenation of olefins catalyzed by this polymer supported complex together with those obtained with the homogeneous (DIOP) Rh^I complex are indicated on Figure 10.

The high optical yields obtained are quite comparable to those observed with the homogeneous catalyst. The same absolute configuration of the products was also observed. It is interesting to note a slower rate than in homogeneous phase probably due to a rate determining diffusion.

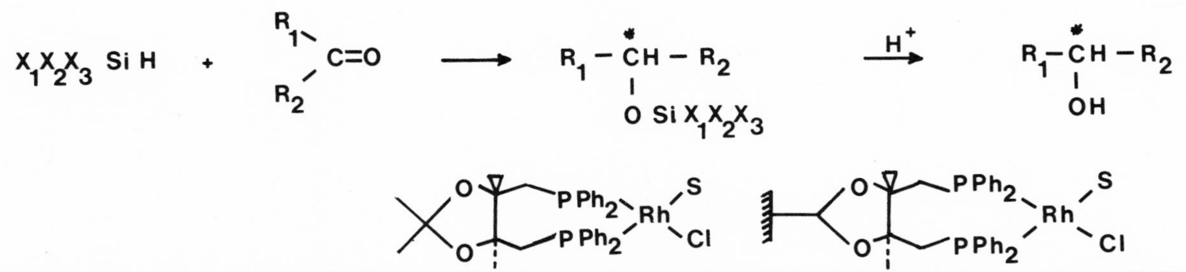

KETONE	SILANE	HOMOGENEOUS CAT.		SUPPORTED CATALYST	
		CONV. (%)	OPT. YIELD (%)	CONV. (%)	OPT. YIELD %)
Ø-CO CH_3	H_2Si(Ø)(naphtyl)	100	58	100	58.5
Ø-CO CH(CH_3)(CH_3)	H_2Si(Ø)(Ø)	100	35	100	28

FIG. 8. Asymmetric hydrosilylation of ketones.

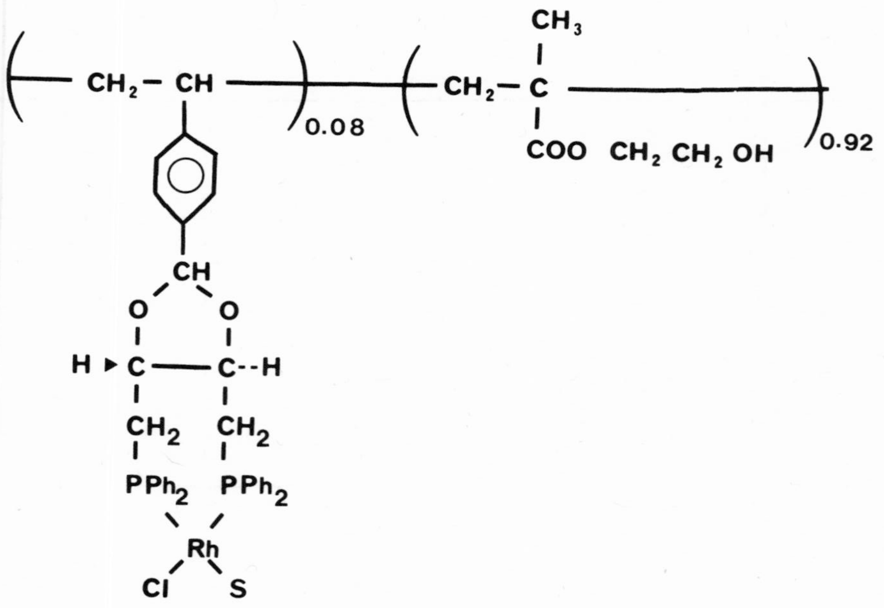

FIG. 9. Catalytic asymetric synthesis via polymer bound chiral ligand.

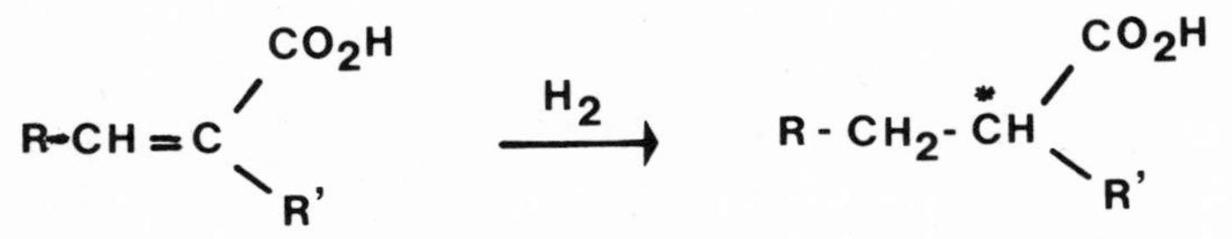

$$R\text{-}CH=C\underset{R'}{\overset{CO_2H}{\diagup}} \xrightarrow{H_2} R\text{-}CH_2\text{-}\overset{*}{CH}\underset{R'}{\overset{CO_2H}{\diagup}}$$

substrate	time (h)	conversion (%)	optical yield supported	optical yield homogen	configuration
R = H , R' = NHCOCH₃	5	100	52	73	R
R = Ph , R' = NHCOCH₃	12	100	86	81	R
R = H , R' = Ph	12	100	58	63	R

FIG. 10. Asymmetric hydrogenation of olefins by polymer supported Rh^I-diop catalyst.

The fact that in hydrosilylation of ketones as well as in hydrogenation of olefins one can obtain the same optical yield as in homogeneous catalysis is a good demonstration that the steric requirements
of the catalytic sites are quite similar in both cases (supported VS
homogeneous).[28]   Even though the rate parameters have been modified
with supported complexes, those results indicate that upon choosing
the right substrate along with the right solvent and the right polarity of the polymer, comparable results must be expected when one
grafts an homogeneous catalyst to a polymeric support. The first
goal of catalysis by supported complexes is thus achieved at least at
the laboratory scale (easy recovery and re-use of the costly catalyst
(or ligand!)).

I.B. Inorganic oxides as a support. A great deal of work[30-35]
has been carried out by British Petroleum which has used intensively
silica as a support for many complexes and which has investigated
a wide range of catalytic reactions. The most interesting aspect of
their work is related to the development of new methods for grafting
many homogeneous catalysts on silica. These methods seem to be quite
general (Fig. 11).

The first method (A) consists to react e.g. $(Et-O)_3Si-(CH_2)_n-PPh_2$
with the silanol groups present on silica to obtain a grafted
phosphine. Unfortunately one does not know exactly the ratio ethoxy/
silane groups which have reacted with the silanol groups, but at least
it seems that the grafted phosphine is strongly linked to the silica
support.[36] It is then possible by ligand exchange to graft any kind
of precursor complex. The second method (B) consists first to exchange the precursor complex with $(EtO)_3-Si(CH_2)_n-PPh_2$. It is then
possible to exchange 1, 2 or more silane ligands before grafting the
complex to the support. These methods which seem to be quite general
present many advantages: there is a wide range of ligand groups
readily prepared (nitrile, amine, pyridine, thiols...). It is possible to vary the distance between the surface and the ligand group. It
is also possible to vary the distance between the ligand groups by
varying the content of silanol groups of the support.[36] The micro
environment of the support may be modified prior to grafting the complex (method A). Finally in the case of method B the coordination
sphere of the complex is known before grafting and one should expect
that it is kept after grafting.

Among the inconvenience of these methods one should mention, as
usual, the lack of information on the real coordination sphere of
the supported complexes (except for carbonyl complexes). Besides,
and this is not always mentioned in the publications, there is sometimes an easy migration of the complex onto the support depending on
many parameters such as the hydrophilic nature of this support, the
rigidity of the ligand, etc. So that in many cases, especially with
the WILKINSON catalyst, metallic particles, which are very highly
dispersed, are formed on the surface. This problem of metal forma-

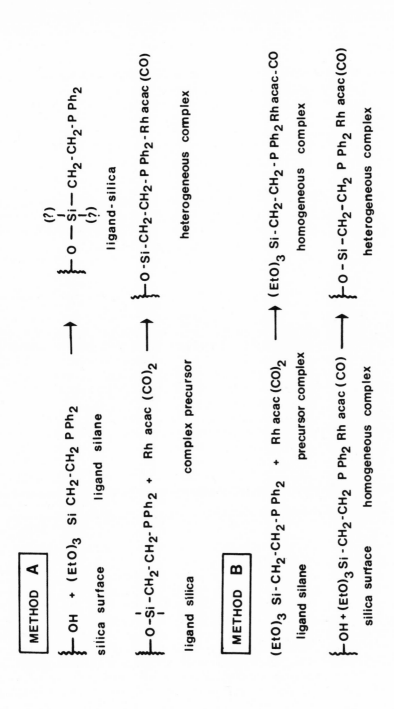

FIG. 11. Grafting of homogeneous catalysis on silica.

tion, which occurs also on polymer supports, has been studied in details by Collman and his group[37] in the case of Rhodium[I] catalysts.
In order to avoid the easy migration of Rh to the support they use
a chelating diphos ligand and the corresponding turnover number is
close to that obtained in homogeneous phase (Fig. 12). Interestingly
is the very high turnover number observed with a typical heterogeneous
catalyst 5% Rh/Al_2O_3 which exhibits a rather different behaviour.
Therefore in the particular case of supported WILKINSON catalyst,
any increased activity observed after grafting must be considered
cautiously. It must be pointed out that metal formation usually
requires an induction period corresponding to the reduction of Rh[I] to
Rh(0).[38]

Another problem concerning supported complexes is related to the
reversibility of ligand dissociation. Stabilization of coordinative
unsaturation is certainly one of the basic aspects of heterogeneous
catalysis; this should also occur with supported complexes provided
the support is rigid enough to prevent coordination of another ligand
far from the metallic center. In some cases stable coordinatively
unsaturated species have been shown to occur with carbonyl complexes
of molybdenum[39] or tungsten[40] supported on alumina or chlorinated
alumina. In other cases the situation has been shown to be more
complicated.[41] Using a phosphinated silica it is quite possible to
obtain with $Ni(CO)_4$ a mono substituted nickle carbonyl on the surface ⧄-L-$Ni(CO)_3$ (Fig. 13). Upon evacuation at 150ºC the carbonyl
ligands have been removed as evidenced by infrared spectroscopy.
The monosubstituted $Ni(CO)_3$L surface compound is instantaneously re-
obtained under CO at room temperature, suggesting that unsaturated
nickel could be still linked to the phosphine ligand. In fact the
situation was more complex since on decarbonylation the nickel mi-
grates onto the support to give under CO, $Ni(CO)_4$ which then reacts
again with the phosphine ligand.

Obviously this problem is also related to the stability of the
supported complexes which seems to depend on many factors as the ratio
ligand/metal (which can in some cases govern the strain of the polymer
matrix) or the chemical affinity of the unsaturated complex toward
the support.

II. The Molecular Nature of the Catalyst Is Lost or Unknown

This field of research which is also related to catalysis by
supported complexes is significantly different from the preceding
one since the method of grafting is such that the support acts as a
ligand directly bonded to the transition metal, so that in many cases
the formation of $Si-O-M_T$ bond or $Al-O-M_T$ bond are involved. It does
not seem that in this case, the molecular nature of the catalyst is
kept at least in the sense of an homogeneous complex, but the resulting catalyst may exhibit very high activity and selectivity compared

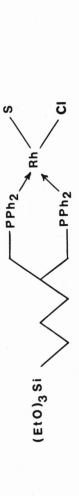

	SUBSTRATE	TURNOVER NUMBER
– HETEROGENOUS CATALYST		
5% Rh / Al$_2$O$_3$	1-decene	5.6
– HOMOGENEOUS CATALYST		
(PPh$_3$)$_n$ Rh Cl n = 2	1-decene	0.002
– SUPPORTED CATALYST		
Si P$_2$ Rh Cl	1-decene	0.0022

FIG. 12. Chelating ligands in order to avoid metal formation.

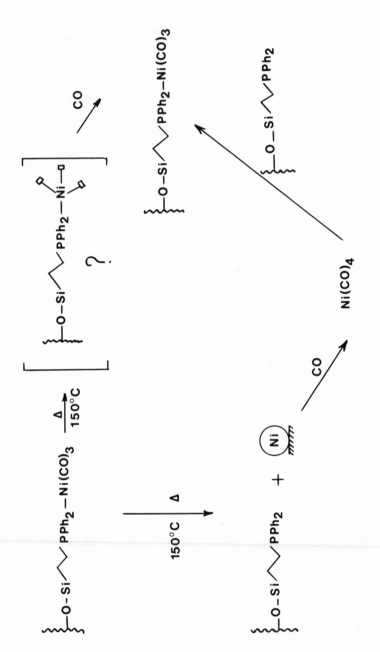

FIG. 13. Ligand dissociation with supported complexes.

with the homogeneous counterpart. Very intense research in this
field is due to Ballard and coworkers at ICI[42-43] and to Yermakov
and coworkers.[44] The general method consists to react an organometallic with the surface groups of alumina or silica and to study
the catalytic properties of this grafted organometallic. In most
cases studied so far the resulting solids exhibit a very high activity
in polymerization or disproportionation of olefins.

Typical examples, given on Figure 14, are related to the reaction
of $(\pi\text{-allyl})_4 Zr$ with silanol groups of a silica surface.[43-44] The
catalytic species assumed to be formed is deduced from the amount of
propoylene evolved by direct interaction of the organometallic with
the surface. The bis-π-allyl structure is then confirmed by destruction with butanol leading again to two molecules of propylene. In
fact the reaction scheme might be more complicated and might depend
strongly, as usual in heterogeneous catalysis, on the water content
of the surface, the presence of strained $Si\!\frown\!O\!\frown\!Si$ bridges a.s.o..
In addition to the simple chemical methods for following these processes, infrared spectroscopy as well as esr[43] (for $(\pi\text{-allyl})_4 Nb$))
have been widely used and they offered a good support in favor of
the proposed structures on the surface which are certainly more defined than many heterogeneous catalysts, prepared in a conventional
way.

The catalytic activity of these supported organometallic compounds have been studied in detail in the case of ethylene polymerization with zirconium complexes.[42] As indicated in Figure 15, the
supported complexes exhibit a higher activity than the homogeneous
ones, the nature of the support (silica or alumina) being also a key
factor. There is no doubt here that the support acts as a ligand
and probably determines an electronic effect on the transition metal
through a Si-O-Zr or Al-O-Zr bond. This was at the origin of the
synthesis of model compounds such as the one represented on Figure 15
with Si-O-Zr bonds. These compounds are good models of what is really
occuring on the surface of the oxide after grafting the organometallic. It is interesting to note the very high activity in polymerization of ethylene on these soluble compounds.

Although these types of model-compounds of the surface of an inorganic oxide have not received so far a great deal of attention, they
might be in the near future a very promising fundamental approach of
heterogeneous catalysis over oxides. Some new results in this field
will appear soon.[45]

The proximity of the transition metal to the surface in these
supported complexes not only implies that electronic effect of the
surface will occur but also implies that steric effect should also
occur, since one can consider that half of the coordination sphere
of the transition metal is masked by the surface. Unusual selectivities should therefore be observed. Such steric effects have been

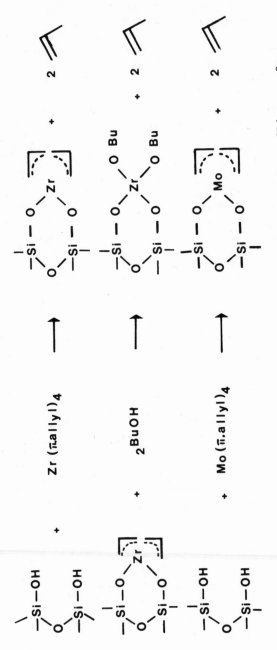

FIG. 14. Direct interaction of organometallic compounds with a silica surface.

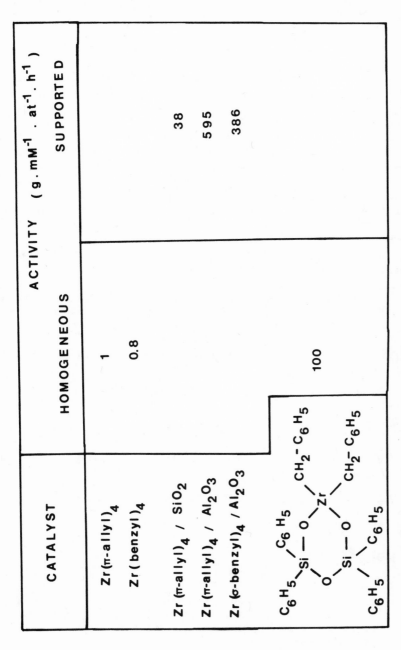

FIG. 15. Polymerisation of ethylene with homogeneous or supported σ-alkyl or π-allyl Zr complexes.

shown to occur in stereospecific polymerization of propylene[46] or in
metathesis of olefins with zerovalent precursor complexes of tungsten[47] activated by Lewis acids in homogeneous phase or supported in
alumina (Fig. 16).

The stereoselectivity in metathesis of cis-2-pentene can be
determined by the trans/cis ratio of 2-butene at 0% conversion. This
stereoselectivity is quite modified in favor of cis isomers with complexes supported on alumina. Obviously the steric effect of the bulky
ligand which is the surface, favors the cis coordination of the olefin
to the metallocarbene moiety.[48]

Another alternative of using these supported complexes has been
also developed by Yermakov,[44] and closely resembles the usual techniques of heterogeneous catalysis. After grafting the organometallics to the surface, drastic reducing treatments under hydrogen or
oxidizing treatments under oxygen will lead to a wide range of oxidation states. The advantage of using these organometallic as starting
material is due to the fact that a high degree of dispersion is obtained on the surface. The general possibilities of these catalysts
are summarized below:

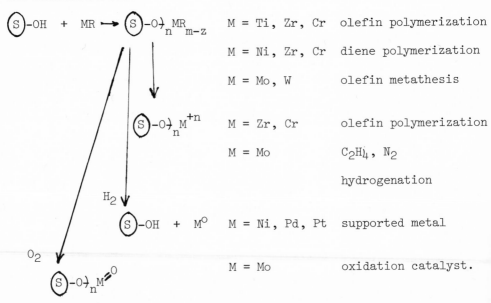

M = Ti, Zr, Cr   olefin polymerization

M = Ni, Zr, Cr   diene polymerization

M = Mo, W   olefin metathesis

M = Zr, Cr   olefin polymerization

M = Mo C_2H_4, N_2

                 hydrogenation

M = Ni, Pd, Pt   supported metal

M = Mo   oxidation catalyst.

As a conclusion, this field of supported organometallic species
is certainly the most promising from an applied point of view. Polymerization processes by Montedison, ICI, Solvay are already working
with this third generation of catalysts. Union Carbide has also
developed a new catalyst for ethylene polymerization, which is obtained by reacting Cp_2Cr (Cp = cyclopentadienyl) with a high surface

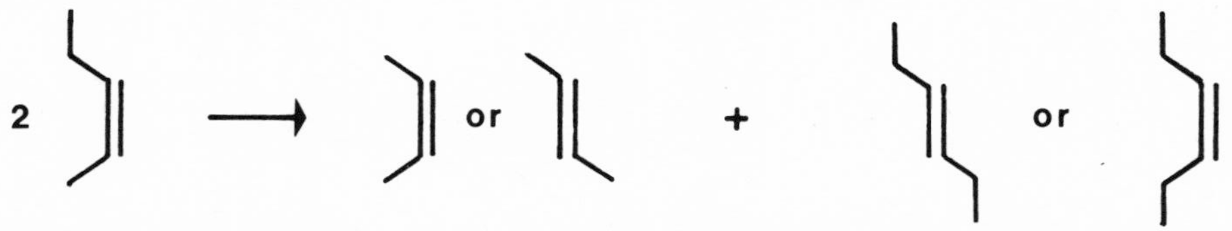

CATALYST	RATIO TRANS/CIS OF C_4 AT 0% CONVERSION	
	HOMOGENEOUS ($+ C_2H_5 AlCl_2, O_2$)	SUPPORTED (on η Al_2O_3)
$W(CO)_6$	0.69	0.37
$W(CO)_5\ PPh_3$	0.73	0.38
$W(CO)_5\ P(n\text{-}bu)_3$	0.76	0.38
$W(CO)_5\ P(OPh)_3$	0.76	0.40

FIG. 16. Olefin metathesis: Steric effect of a surface with supported complexes.

silica.[49] Dimerization of propylene is also achieved when $(\pi\text{-allyl})_2$-
Ni is reacted with the surface groups of a silica gel and then activated by an organoaluminum compound.[50] Nevertheless the obtained
catalysts are certainly less defined than any homogeneous counterpart and obviously much remains to be done before understanding the
nature of the coordination sphere of those species. Nevertheless
it seems that in those systems, as already outlined by Commereuc and
Martino,[3] the support has a triple role: insolubilizing the catalyst,
dispersing the active centers, activating agent or co-catalyst.

It should be also mentioned that clusters supported on inorganic
oxides or on polymers might also exhibit very unusual catalytic properties which might result from the easy transition from a molecular
state (clusters) to the metallic state (small metal particles)[52] with
a rigidity and reactivity of a metal-metal bond which might be higher
than in solution. Some results have already appeared in this field
which is at its very beginning.[53,54,26] In the case of silica supported nickel clusters it seems that the triangular metal-metal framework is kept after decarbonylation, and present interesting catalytic
properties [54] (Fig. 17).

CONCLUSIONS

Although it might seem arbitrary to divide the field of catalysis
by supported complexes into two mains regions, namely the molecular
nature of the catalyst is kept or is lost, this distinction facili-
tates the understanding of a growing field which overlaps so much
homogeneous and heterogeneous catalyst that one gets easily confused.

When one grafts a catalyst on a support with the aim of keeping
its molecular nature, it seems now that any homogeneous catalyst
attached to an insoluble support maintains its basic catalytic activity and selectivity provided the choice of the support is closely related to the catalytic reaction which must be carried out. Besides
additional changes in activity and selectivity may be superimposed
on the basic properties of the starting complex.

When one grafts an organometallic compound on a support without
trying to preserve its molecular nature, one can obtain very high
activity and (or) selectivity compared with the homogeneous counterpart. This has been at the origin of a new class of heterogeneous
catalysts of high industrial importance. The surface is then a real
ligand with specific electronic and steric effects.

In both cases we are facing the same problems as heterogeneous
catalysis, the major one being the lack of information on the coordination sphere of the active catalyst. A more fundamental approach is therefore necessary to define more precisely the coordination sphere of the supported complexes, the kinetic problems of dif-

POLYMER BOUND CLUSTER

$-\!\!\left(P\,Ph_2\right)_n\!\!- Rh_4\,(CO)_{12-n}$

$-\!\!\left(P\,Ph_2\right)_n\!\!- Rh_6\,(CO)_{16-n}$

SUPPORTED CLUSTERS

SILICA SUPPORTED CLUSTER

$Ni_3\,Cp_3\,(CO)_2$ / SiO_2

$Ni_2\,Cp_2\,(CO)_2$ / SiO_2

FIG. 17.

fusion and chemical affinity, the stability of the supported complexes. More generally a great progress could result from a better control and exploitation of the cooperative effect of the support.

Finally it should be pointed out that new developments should compete with the field of catalysis by supported complexes in so far as the recovery of the costly catalyst is required. Membrane systems as well as water soluble catalysts with water soluble ligands might be in the future a solution to this technological problem.[55, 56, 57]

REFERENCES

1. J. Manassen, Catalysis, Progress in research, Basolo and Burwell ed., Plenum Press, London, 177 (1973).
2. N. Kohler and F. Dawans, Rev. Inst. Fr. Pet. 27, 105 (1972).
3. D. Commereuc and G. Martino, Rev. Inst. Fr. Pet., 30, 89 (1975).
4. J. C. Bailar, Cat. Rev., 10, 17 (1974).
5. J. Manassen, Plat. Met. Rev., 15, 142 (1971).
6. Z. M. Michalska and D. E. Webster, Plat. Met. Rev., 18, 65 (1974).
7. G. Jannes, Catalysis heterogeneous and homogeneous, Delmon et Jannes, ed., Elsevier, p. 83 (1975).
8. C. U. Pittman and G. O. Evans, Chem. Tech., 561 (1973).
9. S. L. Davidova and N. A. Plate, Coord. Chem. Rev., 16, 195 (1975).
10. E. Cernia and M. Graziani, J. Appl. Polym. Sci., 18, 2725 (1974).
11. R. H. Grubbs and R. C. Kroll, J. Am. Chem. Soc., 93, 3062 (1971).
12. R. H. Grubbs, R. C. Kroll and E. Sweet, J. Macromol. Sc. Chem. A-7, 1047 (1973).
13. R. H. Grubbs, C. Gibbons, R. C. Kroll and C. Brubaker, J. Am. Chem. Soc., 95, 2373 (1973).
14. C. U. Pittman, G. Evans, R. Jonis and R. Felis, Proc. 6th Int. Conf. Organomet. Chem. Amsterdam (1973).
15. G. Evans, C. U. Pittman, R. Mc. Millan, R. Beach and R. Jones, J. Organomet. Chem., 67, 295 (1974).
16. C. U. Pittman, L. R. Smith and R. M. Hanes, J. Am. Chem. Soc., 97, 1742 (1975).
17. C. U. Pittman, S. E. Jacobson and H. Hiramoto, J. Am. Chem. Soc., 97, 4774 (1975).
18. C. U. Pittman, and R. L. Smith, 97, 1749 (1975).
19. G. Braca, G. Sbrana, C. Carlini and F. Ciardelli, Catalysis, homogeneous and heterogeneous, Delmon and Jannes, p. 307 (1975) and ref. therefrom.
20. M. Graziani, G. Strukul, M. Bonivento, F. Pina, E. Cernia, and N. Palladino, Catalysis homogeneous and heterogeneous, Delmon and Jannes, ed., p. 33 (1975).
21. C. U. Pittman, L. R. Smith and S. E. Jacobson, Catalysis homogeneous and heterogeneous, Delmon and Jannes, ed., p. 393 (1975).
22. R. F. Batchelder, B. C. Gates and F. P. Kuijpers, Prepr. VIth

Int. Cong. Cat. paper A-40, London (1976).

23. G. Bernard, Y. Chauvin and D. Commereuc, Bull. Soc. Chim. Fr.,
1163 (1976).

24. G. Bernard, Y. Chauvin and D. Commereuc, Bull. Soc. Chim. Fr.,
1168 (1976).

25. G. Bernard, thesis, Paris (1977).

26. J. P. Collman, L. S. Hegedus, M. P. Cooke, J. Norton, G. Dulcetti
and D. N. Marquardt, J. Am. Chem. Soc., $\underline{94}$, 1789 (1972).

27. K. Solc, W. H. Stockmayer and W. Gobush, Macromol. $\underline{8}$, 689 (1975).

28. W. Dumont, J. C. Poulin, T. Dang and H. Kagan, J. Am. Chem. Soc.,
$\underline{95}$, 8295 (1973).

29. N. Takaishi, H. Imai, C. A. Bertelo and J. K. Stille, J. Am.
Chem. Soc., $\underline{98}$, 5400 (1975).

30. I. V. Howell, R. D. Hancock, R. C. Pitketly and P. J. Robinson,
Catalysis homogeneous and heterogeneous, Delmon and Jannes, ed., p.
349 (1975).

31. K. G. Allum, R. D. Hancock, S. Mc. Kenzie and R. C. Pitketly,
Proceed. Vth Int. Cong. Cat. Palm Beach, Hightower ed., p. 477 (1973).

32. K. G. Allum, R. D. Hancock, I. W. Howell, R. C. Pitketly and
P. J. Robinson, J. Organomet. Chem., $\underline{87}$, 189 (1975).

33. K. G. Allum, R. D. Hancock, I. W. Howell, S. Mc. Kenzie, S.
Pitketly and R. Robinson, J. Organomet. Chem., $\underline{87}$, 203 (1975).

34. K. G. Allum, R. D. Hancock, I. K. Howell, T. E. Lester, S. Mc.
Kenzie, R. C. Pitketly and P. J. Robinson, J. Catalysis, $\underline{43}$, 331
(1976).

35. K. G. Allum, R. O. Hancock, I. V. Howell, R. C. Pitketly, and
P. J. Robinson, J. Catalysis, $\underline{43}$, 322 (1976).

36. A. K. Smith, J. M. Basset and P. M. Maitlis, Proc. II Franco-
Soviétique Colloque sur la Catalyse, C.N.R.S. Ed., p. 81 (1976).

37. D. N. Marquardt, PhD Thesis, Stanford (1974).

38. J. Conan, M. Bartholin and A. Guyot, J. Mol. Cat., $\underline{1}$, 375 (1976).
M. Bartholin, private communication.

39. A. Brenner and R. L. Burwell, J. Am. Chem. Soc., $\underline{97}$, 2565 (1976).

40. J. M. Basset, Y. Ben Taarit, J. L. Bilhou, J. Bousquet, R. Mutin
and A. Theolier, Preprint VI Int. Cong. Cat. London, p. 47 (1976).

41. A. K. Smith, J. M. Basset and P. Maitlis, J. Mol. Cat. under
press (1977).

42. D. G. H. Ballard, Adv. Cat. $\underline{23}$, 263 (1973) and references therein.

43. J. P. Candlin and H. Thomas, Homogeneous Catalysis II, D. Foster
and J. Roth, ed., Adv. Chem. Ser., $\underline{132}$, 212 (1974).

44. Y. I. Yermakov, Cat. Rev., $\underline{13}$, 77 (1976) and references therein.

45. P. Teyssie, Catalysis heterogeneous and homogeneous, Delmon and
Jannes Ed., p. 289 (1975).

46. G. Natta and I. Pasquon, Adv. Cat. XI, 2(1959).

47. J. M. Basset, J. L. Bilhou, R. Mutin and A. Theolier, J. Am.
Chem. Soc., $\underline{97}$, 7376 (1975).

48. J. L. Bilhou, J. M. Basset, R. Mutin and W. F. Graydon, J. Am.
Chem. Soc., $\underline{97}$, 7376 (1975).

49. Union Carbide Corp. Dutch Pat. Appl. No 6816149 (1969).

50. W. Skupinski and S. Malinowski, J. Organomet. Chem. $\underline{99}$, 465
(1975).

51.   W. Skupinski and S. Malinowski, J. Organomet. Chem., 117, 183 (1976).
52.   J. M. Basset and R. Ugo, Aspects of homogeneous catalysis, R. Ugo, ed., III, 3 (1976).
53.   M. Ichikawa, Chem. Comm., 11 (1976).
54.   M. Ichikawa, Chem. Comm., 26 (1976).
55.   F. Ioo and M. T. Beck, React. Kinet. Cat. Lett., 2, 257 (1975).
56.   J. Manassen, Journées de Catalyse homogéne supportée (Eveux) (1976).
57.   V. Schurig and E. Bayer, Chem. Tech., 6 (3), 212 (1976).

METAL CLUSTERS IN CATALYSIS

Jack R. Norton

Department of Chemistry, Princeton University

Princeton, New Jersey 80540

Large numbers of polynuclear transition-metal complexes have now been characterized by X-ray crystallography.[1] Many of them, particularly the larger aggregates,[1d] have structures resembling those of metals. On the basis of this and other similarities, a number of workers[2] have suggested that useful analogies may be drawn between clusters and metal surfaces.

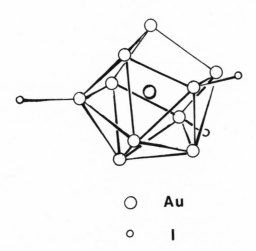

○   **Au**

o   **I**

Figure 1. $Au_{11}I_3L_7$ (phosphine ligands omitted for clarity)

An excellent example of the structural relationship is offered by $Au_{11}X_3L_7$ (Figure 1), where the eleven gold atoms form a truncated icosahedron similar to that found in intermetallic compounds and stoichiometric alloys.[3] An even better and more recent example is the hexagonal close-packed structure of $Rh_{13}(CO)_{24}H_3^{-2}$ (Figure 2).[4] Such hcp structures are common for pure metals occurring, for example, for osmium, ruthenium and, under some conditions, cobalt (curiously, pure rhodium is cubic-close packed instead)[5].

It is possible that the study of these polynuclear complexes or clusters may increase understanding of surface chemistry and hetero-geneous catalysis. More likely to be realized, however, is the hope that such metal clusters will function as "soluble heterogeneous catalysts" and display novel catalytic activity themselves. Ideally, these materials should combine the specificity of present homogeneous catalysts with the range of activity known for metal surfaces.

There have been a number of attempts to employ polynuclear metal

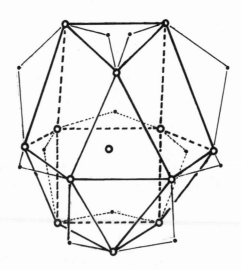

o Rh   terminal CO not shown

• CO bridge

Figure 2. $\left[Rh_{13}(CO)_{24}H_3\right]^{-2}$

complexes as catalysts. Consideration of the results will illustrate
both the possibilities and the difficulties of such uses for these
materials. Some time ago, BASF reported the reaction of propylene
with carbon monoxide in water to form butanols in the presence of
$HFe_3(CO)_{11}^-$ [6]. At the temperatures employed, however, it seems likely
that many mononuclear fragments are present and that they are the
catalytically active species; evidence to this effect has recently
been presented.[7] $Ru_3(CO)_{12}$, and compounds which generate it under the
reaction conditions, have been shown to catalyze the reduction of nitrobenzene to aniline by carbon monoxide and hydrogen.[8] Also derivatives of $Rh_4(CO)_{12}$ have been reported to effect olefin hydroformylation at atmospheric pressure and room temperature.[9] In both of the
last two reactions the nuclearity of the active species is unclear.

A very interesting cluster-based $(Rh_6(CO)_{16})$ system catalyzes
the oxidation of cyclohexanone to adipic acid.[10] However, on the
basis of the internal evidence given in the paper[10] it seems certain
that cluster fragmentation occurs under the reaction conditions.

The most exciting development has undoubtedly been the report[11]
that various rhodium complexes catalyze the formation of ethylene glycol and related compounds directly from carbon monoxide and hydrogen.
Although the pressures involved are extremely high, it has proven
possible to find infrared evidence for the presence of polynuclear
rhodium carbonyl anions under the reaction conditions;[12] this observation is not surprising in view of the known chemistry of these anions.[13] It will be quite difficult, however, to learn the identity of
the catalytically active species and the mechanism of this reaction.

A related reaction which also poses formidable mechanistic difficulties is the recent formation of methane and ethane from carbon
monoxide and hydrogen at low pressures.[14] Clusters used are
$Ir_4(CO)_{12}$, $Os_3(CO)_{12}$, and their derivatives. Conversion is low,
turnovers are small, and rates are too slow to make mechanistic studies practical.

Difficulties in learning the nature of the catalytically active
species are quite general with clusters. The polynuclear complex
originally introduced or generated, even though present in overwhelming excess, may not be the catalytically active species, and
the fact that a given catalysis product is not produced with known
mononuclear species is not evidence for true polynuclear catalysis, as
the active catalyst may be a previously unknown mononuclear species
generated in situ by degradation of the cluster.

A classic example of this difficulty is offered by the early
claim[15] that the dimerization of norbornadiene by $Zn[Co(CO)_4]_2$ represented polynuclear catalysis, as the product, so-called Binor-S,
had not been observed with previous mononuclear catalysts. Subsequently, Schrock and Osborn, however,[16] produced Binor-S with mono-

nuclear $(Rh(norbornadiene)_2L)^+$ as catalyst, and it is now generally
accepted that the use of Lewis acid cocatalysts with mononuclear
dimerization catalysts promotes Binor-S formation.[17]

In only one case has the mechanism of a cluster-catalyzed reaction been worked out in detail. Shapley[18] and Deeming[19] have examined the isomerization and hydrogenation of olefins in the presence
of $H_2Os_3(CO)_{10}$, and, by kinetic studies and isolation of stabilized
analogs of proposed intermediates, have established the essential
validity of the scheme shown in Figure 3. The rate of isomerization
is first-order in $H_2Os_3(CO)_{10}$ and first-order in alkene, and largely

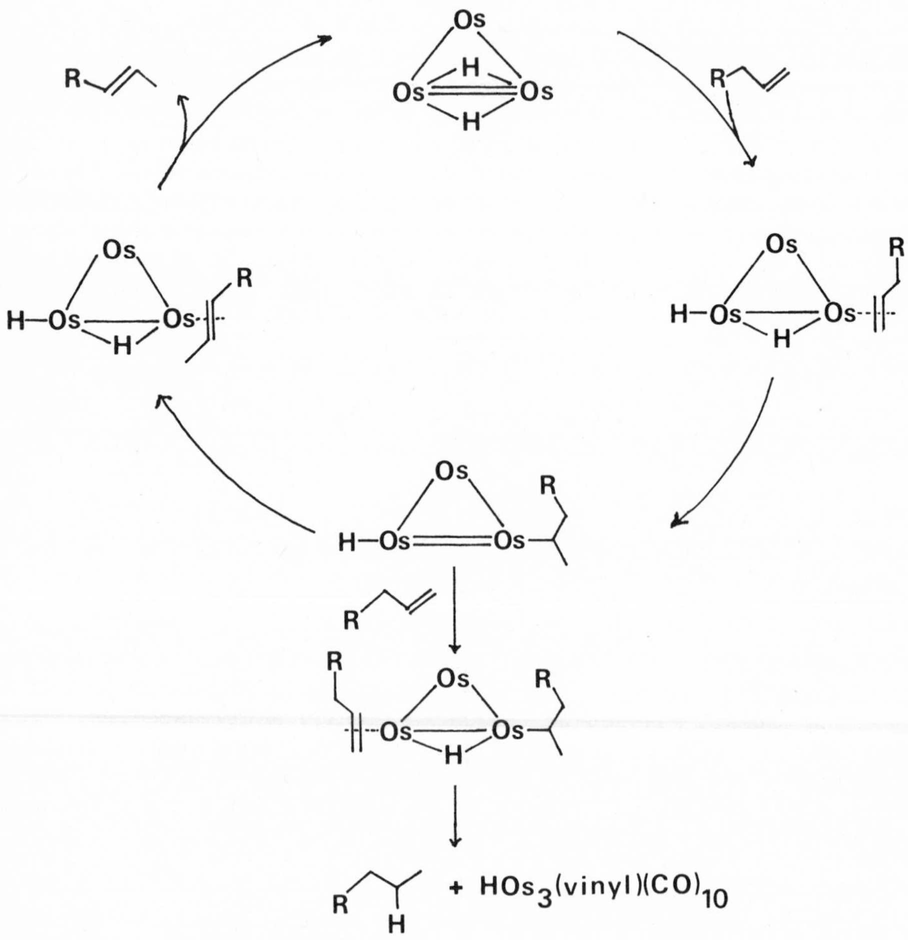

Figure 3. Scheme for Catalysis of Olefin Isomerization and Hydro-
           genation by $H_2Os_3(CO)_{10}$

independent of the nature of R; apparently either the olefin coordination or the olefin insertion step is rate-determining. The ease
with which the species which have formal Os-Os double bonds in Figure
3 can coordinate additional ligands suggests that metal-metal multiple bonds are the cluster equivalent of vacant coordination sites in
mononuclear homogeneous catalysis. Finally, the isolation of hydrido-
alkenyl and stabilized hydridoalkyl clusters analogous to those shown
in Figure 3 makes it unarguable that in this catalytic system the
polynuclear framework does remain intact.

It is important to note that this conclusion did not necessarily
follow from the reaction kinetics alone. Such studies at best define
the composition of the transition state for the rate-determining step,
and do not in general show whether or not the metals remain bonded
during that step and during subsequent fast steps.

Of course, it can be argued that it is precisely the ability
to dissociate readily into highly reactive coordinatively unsaturated
mononuclear fragments that will make polynuclear metal complexes
effective as homogeneous catalysts. Support for this view is offered
by the work of Muetterties on $Ni_4(CNC(CH_3)_3)_7$ (Figure 4).[20] Reference
to the known chemistry of related palladium and platinum complexes
suggests that this type of compound dissociates readily into mononuclear fragments. Such a process probably explains the fact that
this nickel cluster, unlike mononuclear $Ni(CNC(CH_3)_3)_4$, can catalyze

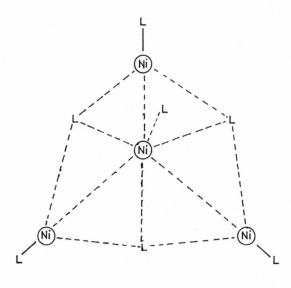

$$L = CNC(CH_3)_3$$

Figure 4. $Ni_4[CNC(CH_3)_3]_7$

cyclization processes at room temperature. Acetylene is converted to benzene, and butadiene to various cyclo-octadienes.[20] Acetylenes can be selectively hydrogenated to <u>cis</u> olefins.[21]

However, if the uses of polynuclear complexes are confined to their serving as precursors for catalytically active fragment complexes, the results will not differ greatly from those obtainable with known mononuclear homogeneous catalysts. For example, no mononuclear complex has demonstrated the ability to effect the reduction of carbon monoxide by hydrogen, and it is likely that only a polynuclear framework can catalyze this process.

There are certainly properties of heterogeneous catalyst surfaces which cannot possibly be duplicated by mononuclear homogeneous catalysts. They are:
1) <u>multiple binding of a single substrate</u>, resulting in precise control of reactant geometries and consequent rate accelerations;
2) the <u>binding of different substrate fragments to adjacent metal atoms</u>;
3) <u>migration of bound molecules on</u> (or through) <u>catalyst surfaces</u>, permitting substrates to react regardless of their initial binding site;
4) <u>chemically significant interactions among various metal atoms at the substrate binding site</u>.
It is thus important to discover whether particular polynuclear metal complexes possess these properties.

A large number of polynuclear complexes are known which contain

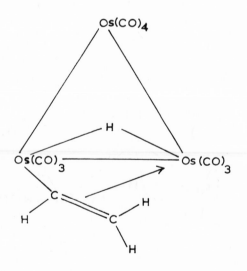

Figure 5. $HOs_3(CO)_{10}(CHCH_2)$

organic ligands -- potential substrates -- bound to more than one metal atom. Examples of such <u>multiple binding</u> -- the first property we seek -- are the bridging vinyl group in $HOs_3(CO)_{10}(CHCH_2)$ (Figure 5)[22] and the ethylidyne group in $H_3Ru_3CO_9CCH_3$ (Figure 6).[23] Numerous additional examples could be given.

Recently we have prepared[24] a number of compounds which offer examples of the second property mentioned above: the <u>binding of substrate fragments to adjacent metal atoms</u>. If <u>cis</u>-$Os(CO)_4(H)CH_3$ is allowed to stand in a sealed tube for a day at room temperature, the principal products are (initially) $HOs(CO)_4Os(CO)_4CH_3$ and (later) $Os_3(CO)_{12}(CH_3)_2$. The initial product, the dinuclear methylhydride $HOs(CO)_4Os(CO)_4CH_3$, is a colorless liquid unstable to air and light; it is, however, fairly stable thermally, evolving methane and CO slowly above 70° to give a variety of yet-uncharacterized products. For better characterization it was treated with CCl_4 to give the expected derivative $ClOs(CO)_4Os(CO)_4CH_3$, a stable white crystalline solid.

The structures assigned to these complexes by spectroscopic means are shown in Figures 7 and 8. $Os_3(CO)_{12}(CH_3)_2$, for example, shows only one methyl peak in the nmr (even when a superconducting magnet is employed) and has too many carbonyl peaks in the ir for a symmetric structure with methyl groups lying along the triosmium axis. The structure shown in Figure 8 -- methyl <u>cis</u> to the triosmium axis -- is further supported by that found by X-ray analysis[25] for the related (the carbonyl region ir spectra of the two species are almost superimposable) molecule $Os_3(CO)_{12}I_2$.

Although these compounds surely exist in solution as a variety of rotamers, Figures 7 and 8 are drawn in the forms which best illustrate the cluster-surface analogy. $HOs(CO)_4Os(CO)_4CH_3$ and and $ClOs(CO)_4Os(CO)_4CH_3$ suggest methane and methyl chloride, respectively, chemisorbed on a "surface" of two osmium atoms -- here stabilized by carbonyl ligands instead of being part of a larger array of

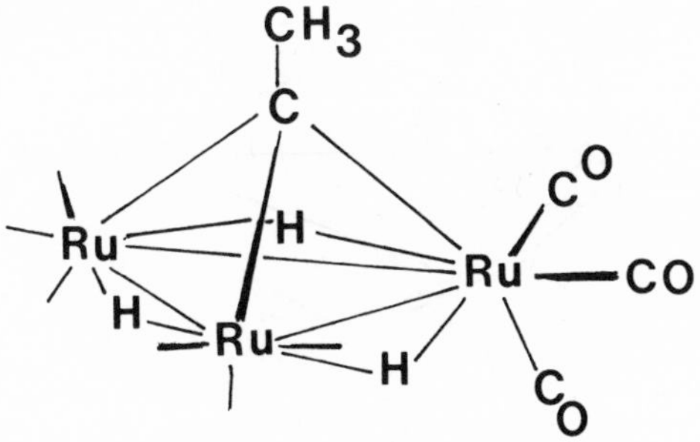

Figure 6. $H_3Ru_3(CO)_9CCH_3$

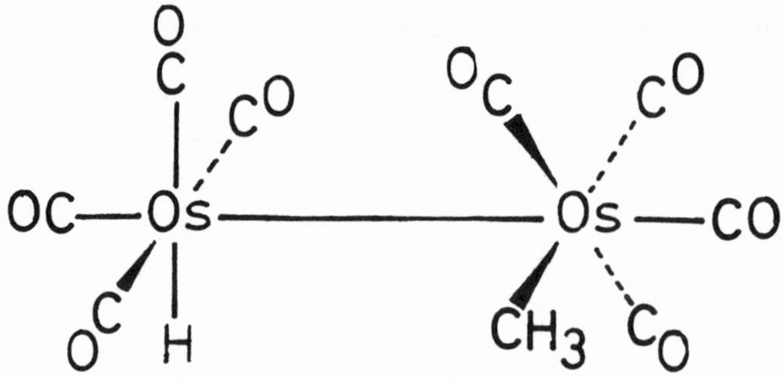

Figure 7. $HOs(CO)_4Os(CO)_4CH_3$ (the structure of $ClOs(CO)_4Os(CO)_4CH_3$ is analogous)

metal atoms. $Os_3(CO)_{12}(CH_3)_2$, although not analogous to a chemisorbed species, represents two methyl groups on nearby osmium atoms of a stabilized "surface."

The third surface property, <u>migration of bound species</u>, has been observed in a large number of polynuclear complexes. The particular ligands most amenable to study, and therefore most widely investigated, are carbonyl and hydride. Rhodium systems are particularly attractive for study, as the rhodium nucleus is 100% spin 1/2. For example, the bridging and terminal carbonyl groups of the simple dimer $(\pi-C_5H_5)_2Rh_2(CO)_3$ are readily identifiable in the low temperature carbon-13 NMR spectrum. At room temperature rearrangement is rapid enough to make all carbonyl ligands equivalent.[26] Later work has produced similar results for Rh_4CO_{12}[27] and a large number of polynuclear rhodium anions of the sort implicated above

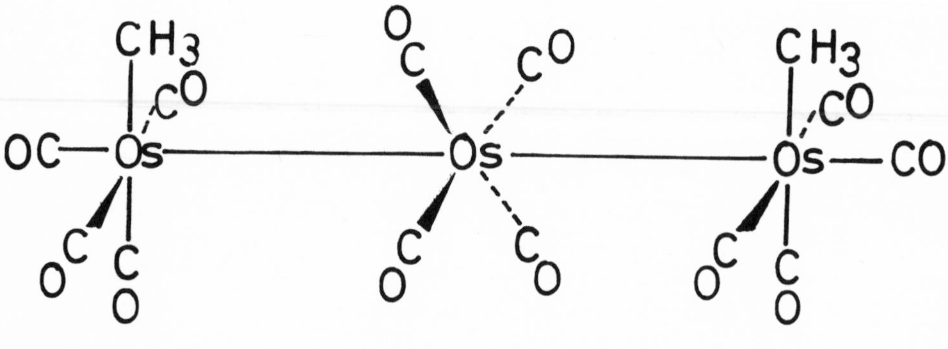

Figure 8. $Os_3(CO)_{12}(CH_3)_2$

as catalytically active species.[28]   There have also been extensive studies of the migration of hydride ligands on clusters such as $[H_3Ru_4(CO)_{12}]^-$.[29]

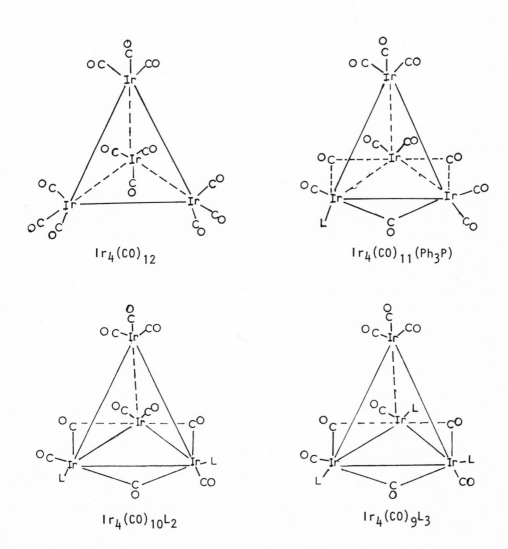

Figure 9. $Ir_4(CO)_{12}$ and Substituted Derivatives

The fourth and remaining property we seek among polynuclear
metal complexes is a <u>chemically significant interaction among metals
at the binding site</u>. This property is demonstrated by sequential
substitution rate studies on the iridium clusters I-IV (Figure 9).[30]
Note that a lack of interaction would be reflected in identical
specific rates for successive substitution reactions at different
metal atoms. Such a result was, in fact, obtained for consecutive
substitutions on the two ruthenium atoms in $Ru_3(CO)_{10}(NO)_2$ not
connected by a metal-metal bond.[31]

The results of such studies on the IR_4 system are given in Table
1. Each step is faster than the succeeding one, and thus regardless
of starting material, the product observed is IV. The activation
parameters for initial substitution on I imply an associative process; the observed rate constant is dependent on the nature and concentration of phosphine as expected for a reaction second order overall. The activation parameters for substitution on II and III, as
well as the observed independence of concentration in nature of
entering phosphine, correspond well to the observed first order rate
law and to a classic dissociative mechanism.

Of course, carbonyl dissociation from (I) must occur at a finite
rate even though its contribution to the phosphine substitution rate
is much less than that of the observed second order term. The maximum possible contribution from a dissociative path is $4 \times 10^{-7} sec^{-1}$
at $75°$. Thus at this temperature the increase in the rate of carbonyl ligand dissociation from unsubstituted (I) to disubstituted
(III) is a factor of 42,000. An apparently minor structural change,
from non-bridged I to the bridged structure seen subsequently, thus
transmits an enormous increase in the carbonyl ligand dissociation
rate. The interactions among the iridium atoms connected by metal-
metal bonds are thus indeed significant.

We now consider how to defect genuine polynuclear catalysis when
it occurs in solution. One solution to this problem would be the
use of a properly designed chiral polynuclear framework. If the
chirality is a property only of the intact, metal-metal bonded
framework and not of any conceivable fragment, a non-zero optical
yield in a catalytic reaction will demonstrate that genuine polynuclear catalysis by a metal-metal bonded species has occurred.

Pursuit of this strategy will require considerable knowledge
of chirality in polynuclear metal-metal-bonded complexes. Unfortunately, very few known clusters are chiral, and of these even fewer
are amenable to measurement of racemization barriers and to ultimate
resolution.

Our attention was therefore caught by the structure reported[32]
for $Rh[Fe(PPh_2)(CO)_2(\pi-C_5H_4CH_3)]^+$. This cluster cation has only C_2
symmetry and is thus chiral. Furthermore, the low symmetry is

TABLE 1.

Reaction	Rate Law	Rate Constants	Obsd Specific Rates x 10^6 sec^{-1}, 75°	Relative Specific Rates (75°)	kcal/mol	$\Delta S^{\ddagger}$ e.u.
I → II	k_1 I PPh$_3$	5.6 x 10^{-3} M^{-1} sec^{-1}	1.5	1	20.6	-22
II → III	k_2 II	8.8 x 10^{-5} sec^{-1}	44	30	31.8	+14
III → IV	k_3 III	1.4 x 10^{-3} sec^{-1}	1380	920	31.0	+17

inherent in the Fe-P-Rh-P-Fe framework.

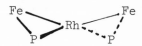

A general synthesis of this type of heteronuclear cationic cluster has been developed (Fp ≡ $(\pi-C_5H_5)Fe(CO)_2$, A = phenyl or p-tolyl,
M = Rh or Ir).[33]

$$FpCl \xrightarrow[C_6H_6]{PAr_2H} \xrightarrow{KPF_6} Fp(PAr_2H)^+PF_6^-$$

$$Fp(PAr_2H)^+PF_6^- \xrightarrow[acetone]{1,8-bis(dimethylamino)naphthalene} FpPAr_2$$

$$FpPAr_2 \xrightarrow[acetone]{Ir(p-toluidine)(CO)_2Cl \text{ or } Rh(CO)_2Cl_2} (FpPAr_2)_2M(CO)Cl$$

$$(FpPAr_2)_2M(CO)Cl \xrightarrow[THF]{AgBF_4} M[Fe(PAr_2)(CO)_2(\pi-C_5H_5)]_2^+BF_4^-$$

The racemization barrier can be measured[33] by variable temperature proton nmr of the p-tolyl derivatives. In these molecules as

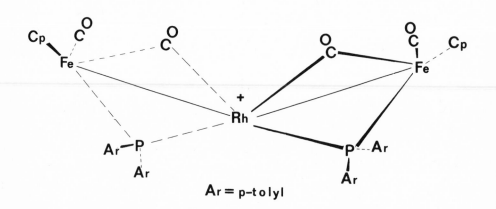

Ar = p-tolyl

Figure 10.   Rh[Fe(P(p-tolyl)$_2$)(CO)$_2$(C$_5$H$_5$)]$_2^+$BF$_4^-$

drawn in Figure 10 there are two inequivalent pairs of aryl groups, the two in each pair being related by the C_2 axis; racemization makes all four aryl groups equivalent. For M = Rh in CD_2Cl_2 the p-tolyl methyl resonance cannot be resolved into two separate signals, but analysis of the ${}^{31}P$ ${}^1H$ spectrum of the aromatic ring (which co-alesces into a single line at -20°, Figure 11) yields a value of 12.6 kcal/mole for the barrier to racemization. For M = Ir the p-tolyl methyl resonance in $ClCD_2CD_2Cl$ gives two signals 4.5 Hz apart, which coalesce at 50°; on this basis, as well as on the basis of the ${}^{31}P{}^1H$ spectrum of the aromatic rings, the racemization barrier of the cation with M = Ir is 17.5 kcal/mole.

A reasonable proposal for the nature of the racemization process in these systems can be based on the observation that, for M = Rh at least, the barrier to bridge-terminal carbonyl exchange (as measured

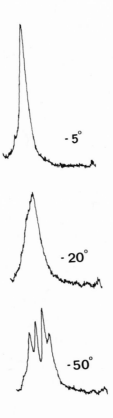

Figure 11. ${}^{31}P$ ${}^1H$ NMR of Aromatic Region of Rh Fe(P(p-tolyl)$_2$)(CO)$_2$(C$_5$H$_5$)$_2$ ${}^+$BF$_4$${}^-$

by ^{13}C nmr) is identical (12.6 kcal/mole) with the barrier to race-mization. An intermediate achiral structure such as

$$(\pi\text{-}C_5H_5)\overset{\overset{\displaystyle O}{\overset{\displaystyle \|}{C}}}{\underset{\underset{\displaystyle O}{\underset{\displaystyle \|}{C}}}{Fe}} \text{---} \overset{\overset{\displaystyle Ar}{|}}{\underset{\underset{\displaystyle Ar}{|}}{P}} \text{---} \overset{\overset{\displaystyle S}{|}}{\underset{\underset{\displaystyle S}{|}}{M^{+}}} \text{---} \overset{\overset{\displaystyle Ar}{|}}{\underset{\underset{\displaystyle Ar}{|}}{P}} \text{---} \overset{\overset{\displaystyle O}{\overset{\displaystyle \|}{C}}}{\underset{\underset{\displaystyle O}{\underset{\displaystyle \|}{C}}}{Fe}}(\pi\text{-}C_5H_5)$$

(S = solvent) will render all aryl groups as well as carbonyls equi-valent. Such structures are observed as the dominant form for com-pounds of this general structure in strong donor solvents.[34] It therefore seems reasonable, and comparison of the results with M = Rh and M = Ir supports this suggestion, that increases in the energy of metal-metal interaction will increase the barrier to racemization. We are thus currently undertaking the synthesis of analogous com-pounds with iron replaced by ruthenium.

Entirely aside from the fact that they are chiral, these mole-cules are attractive as potential heteronuclear catalytic clusters. (Such systems will be able to make increased use of the advantages given above as available to polynuclear homogeneous catalysis in general.) Preliminary studies show that the rhodium-iron cation catalyzes the dimerization of norbornadiene, and that the iridium-iron cation activates hydrogen. Under one atmosphere of H_2 the latter compound forms a thermally unstable burgundy-colored adduct, with a hydride resonance at 32.9τ; the signal is split into a triplet by the phosphorous nuclei, suggesting that the cluster is still in-tact.

References

1. For some recent reviews see a) B. R. Penfold, Perspectives in Structural Chmeistry, 2, 71 (1968); b) R. B. King, Progress in Inorganic Chemistry, 15, 287 (1972); c) J. Lewis and B. F. G. Johnson, Pure Appl. Chem., 44, 43 (1975); d) P. Chini, G. Longoni and V. G. Albano, Adv. Organomet. Chem., 14, 285 (1976).

2. E. L. Muetterties, Bull. Soc. Chim. Belg., 84, 959 (1975); b) R. Ugo, Catal. Rev., 11, 225 (1975); c) E. L. Muetterties, Bull. Soc. Chim. Belg., 85, 451 (1976); d) J. M. Basset and R. Ugo, in Aspects of Homogeneous Catalysis, Vol. III, R. Ugo, ed., Reidel (Holland) (1976).

3. P. Bellon, M. Manassero and M. Sansoni, J. C. S. Dalton, 1481 (1972).

4. V. G. Albano, A. Ceriotti, P. Chini, G. Ciani, S. Martinengo and W. M. Anker, Chem. Commun., 859 (1975).

5. F. A. Cotton and G. Wilkinson, "Advanced Inorganic Chemistry," Third Edition, Interscience, 1972, p. 65.

6. N. Kutepow and H. Kindler, Angew. Chem., 72, 802 (1960).

7. F. Wada and T. Masuda, Chem. Lett., 197 (1974).

8. F. L'Epplattenier, P. Matthys and F. Calderazzo, Inorg. Chem., 9, 342 (1970).

9. P. Chini, S. Martinengo and G. Garlaschelli, Chem. Commun., 709 (1972), and references therein.

10. G. D. Mercer, J. S. Shu, T. B. Rauchfuss and D. M. Roundhill, J. Amer. Chem. Soc., 97, 1967 (1975).

11. R. L. Pruett and W. E. Walker, U. S. Patent Application 210, 538; German Offen. 2,262,318, June 28, 1973; Chem. Abs., 79, 78088k (1973).

12. W. E. Walker, J. B. Cropley and R. E. Pruett, Netherlands patents 7,407,383 and 7,407,412.

13. P. Chini, S. Martinengo and G. Ciordano, Gazz. Chim. Ital., 102, 330 (1972).

14. M. G. Thomas, B. F. Beier and E. L. Muetterties, J. Amer. Chem. Soc., 98, 1296 (1976).

15. G. N. Schrauzer, B. N. Bastian and G. A. Fosselius, J. Amer. Chem. Soc., 88, 4890 (1966).

16. R. R. Schrock and J. A. Osborn, J. Amer. Chem. Soc., 93, 3089 (1971).

17. G. N. Schrauzer, R. K. Y. Ho and G. Schlesinger, Tetrahedron Lett., 543 (1970).

18. J. B. Keister and J. R. Shapley, J. Amer. Chem. Soc., 98, 1056 (1976), and references therein.

19. A. J. Deeming and S. Hasso, J. Organomet. Chem., 114. 313 (1976), and references therein.

20. V. W. Day, R. O. Day, J. S. Kristoff, F. J. Hirsekorn and E. L. Muetterties, J. Amer. Chem. Soc., 97, 2571 (1975).

21. M. G. Thomas, E. L. Muetterties, R. O. Day and V. W. Day, J. Amer. Chem. Soc., 98, 4645 (1976).

22. J. B. Keister and J. R. Shapley, J. Organomet. Chem., 85, C29 (1975).

23. A. J. Canty, B. F. G. Johnson, J. Lewis and J. R. Norton, Chem. Commun., 1331 (1972).

24. J. Evans, S. J. Okrasinski, A. J. Pribula and J. R. Norton, J. Amer. Chem. Soc., 98, 4000 (1976).

25. P. Woodward, private communication.

26. J. Evans, B. F. G. Johnson, J. Lewis and J. R. Norton, Chem. Commun., 79 (1973).

27. a) F. A. Cotton, L. Kracznski, B. L. Shapiro and L. F. Johnson, J. Amer. Chem. Soc., 94, 6191 (1972); b) J. Evans, B. F. G. Johnson, J. Lewis, J. R. Norton and F. A. Cotton, Chem. Commun., 807 (1973).

28. P. Chini, S. Martinengo, D. J. A. McCaffrey and B. T. Heaton, Chem. Commun., 310 (1974).

29. J. W. Koepke, J. R. Johnson, S. A. R. Knox, and H. D. Kaesz, J. Amer. Chem. Soc., 97, 3947 (1975).

30. K. J. Karel and J. R. Norton, J. Amer. Chem. Soc., 96, 6812 (1974).

31. J. R. Norton and J. P. Collman, Inorg. Chem., 12, 476 (1973).

32. a) R. J. Haines, R. Mason, J. A. Zubieta, and C. R. Nolte,

Chem. Commun., 990 (1972); b) R. Mason and J. A. Zubieta, J. Organomet. Chem., 66, 279 (1974).
33. A. Agapiou, S. E. Pedersen, L. A. Zyzyck, and J. R. Norton, manuscript submitted for publication.
34. R. J. Haines, J. C. Burkett-St. Laurent, and C. R. Nolte, J. Organomet. Chem., 104, C27 (1976).

PHASE-TRANSFER CATALYSIS

Luigi Cassar

Montedison SpA, Istituto Ricerche Donegani

Via G. Fauser, 28100 Novara, Italy

Reactions between substances located partly in an organic phase and partly in an aqueous phase are frequently slow and ineffective.

Attempts are therefore frequently made to obtain all the components in homogeneous phase. This limitation has been, in many cases, very difficult to overcome without losing some of the power of the synthetic methods or employing particular and expensive solvents, in order to have all components in solution.

The past few years have seen the growth of a new technique in which reactions are conducted in aqueous/organic or solid/organic two phase systems in the presence of ammonium or phosphonium salts[1] or a complexing agent such as crown ethers or cryptates.[2]

The reactions carried out in these two-phase systems are usually referred to as "phase-transfer catalyzed" according to the term introduced by Starks.[1b]

This new synthetic method has been mainly applied to reactions involving organic and inorganic anions.

The reactions of organic and inorganic anions with organic compounds, of great value in organic synthesis, have been largely studied from the point of view of the influence of solvents on the reactivity of anions. The factors which determine the activity of an anion include anion-cation interaction, aggregation, anion-solvent interaction. In general the activity of an anion can be enhanced under conditions where such interactions are eliminated.

From the practical point of view, a first solution to this problem has been the use of dipolar aprotic solvents which are effective in solvating cations.[3] However there are considerable practical disadvantages connected with the use of these solvents.

Recently the use of two phase-system has been clearly shown to enhance the reactivity of organic anions.[1,2] This technique is based on the introduction of anions in non-polar media in form of their salts with organic cations of sufficient size and lipophilicity. Under these conditions anions are poorly solvated and therefore they exhibit high activity. The onium salts or the complexing agents allow the anions to pass into the organic phase where they react at a high rate with the organic substrate.

Tetraalkylammonium and tetraalkylphosphonium cations as well as complexing agents of alkaline metal cations (crown ethers, cryptates) can be conveniently used.

Serious practical disadvantages are associated with the use of such salts in stoichiometric quantity. However this disadvantage is eliminated by application of two-phase systems (phase transfer catalysis) in which the onium salts or complexing agents are used in catalytic amounts with respect to the reactants.

In this technique the organic reactants, neat or in nonpolar solvents, constitute the organic phase whereas inorganic salts are used in solid state or in form of aqueous solution.

Typical examples of the phase transfer catalysis are:

1. Alkylation of benzyl cyanide in the presence of aqueous sodium hydroxide and a small amount of ammonium salt:[1,4]

$$\text{C}_6\text{H}_5\text{-CH}_2\text{-CN} + \text{C}_2\text{H}_5\text{Br} + \text{NaOH} \xrightarrow[\text{NR'}_4\text{X}]{\substack{\text{NaOH/Benzene} \\ \text{aqueous}}}$$

$$\xrightarrow{\hspace{2cm}} \text{C}_6\text{H}_5\text{-CH-CN} + \text{NaBr} + \text{H}_2\text{O}$$
$$\underset{\text{C}_2\text{H}_5}{|}$$

For this reaction it is otherwise necessary to use anhydrous conditions and a base such as sodium hydride or sodium amide.[5]

2. Reaction of 1-chlorooctane with aqueous sodium cyanide:

$$\text{C}_8\text{H}_{17}\text{Cl} + \text{NaCN} \xrightarrow[\text{NR}_4\text{X (or PR}_4\text{X)}]{\text{NaCN Aqueous}} \text{C}_8\text{H}_{17}\text{CN} + \text{NaCl}$$

This latter reaction can be accomplished in about 2 hours, in 99% yield if 1.3 mole % of tributyl(hexadecyl)phosphonium bromide is present.[1b] Without the catalyst no 1-cyanooctane is formed, even after two weeks boiling.

The onium salt acts as a catalyst because quaternary cations in a non polar environment preferentially associate with the CN^- in respect to Cl^-. This selectivity is important since it dictates which anion will predominate in the organic phase.

3. Formation of cyclopropene derivatives via dichlorocarbene generated from chloroform, concentrated sodium hydroxide and ammonium salt.[1, 5]

The method is extremely simple: stirring of chloroform, an alkene and an excess of concentrated aqueous NaOH in the presence of 1-2 molar percent of some tetraalkylammonium salt results usually in a mild exothermic reaction; dichlorocyclopropane derivative is then isolated in high yield.[1, 5]

Other reactions such as hydrolyses, condensations and oxidations have been executed in two phase systems with catalysts. Some work has been carried out to elucidate the mechanisms of action of the ammonium or phosphonium salts. The possible mechanisms for the three reactions described above are illustrated in Figures 1, 2, 3.

Mechanistic studies have afforded the following evidence which supports the proposed scheme. Small ions (e.g. tetramethylammonium) are not suitable as catalyst, while large ions (e.g. tetrabutylammonium and tetradodecilammonium) are very effective.

The mixing speed has no influence on the reaction kinetics once a limiting value for thorough mixing has been reached.

The rate constant for the alkyl halide/cyanide reaction is proportional to the amount of catalyst and of first order in alkyl halides.

Some mechanistic aspects of these systems are still to be studied even though the main aspects have been clarified.

This new technique has been largely applied in organic syntheses and over 300 papers have been published in the past decade.

The main advantages of this technique are:

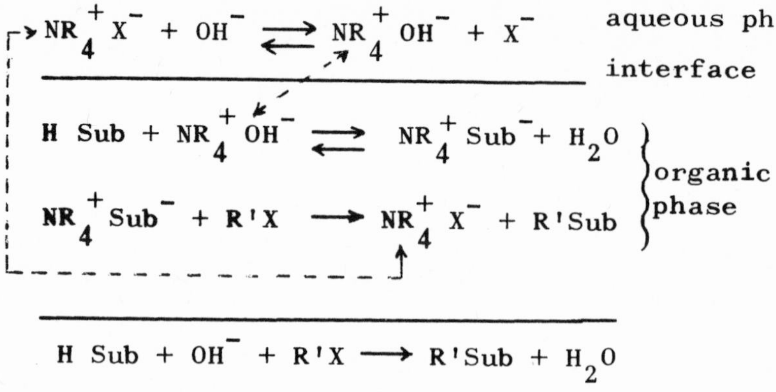

$$NR_4^+ X^- + OH^- \rightleftharpoons NR_4^+ OH^- + X^- \qquad \text{aqueous phase}$$

interface

$$H \ Sub + NR_4^+ OH^- \rightleftharpoons NR_4^+ Sub^- + H_2O$$

$$NR_4^+ Sub^- + R'X \longrightarrow NR_4^+ X^- + R'Sub$$

organic phase

$$H \ Sub + OH^- + R'X \longrightarrow R'Sub + H_2O$$

Fig. 1.   Course of phase-transfer alkylation.

1.   It is possible to run reactions between reactants having no common solvents.

2.   Reactions can be carried out without organic solvents.

3.   Enhancement of the reaction rates of anions in non-polar media.

4.   Inorganic anions produced during the reaction are removed from the organic phase.

5.   Use of NaOH instead of dangerous and moisture sensitive reagents such as NaH, $NaOCH_3$, $NaNH_2$ etc.

6.   Highly selective transformations can be achieved.

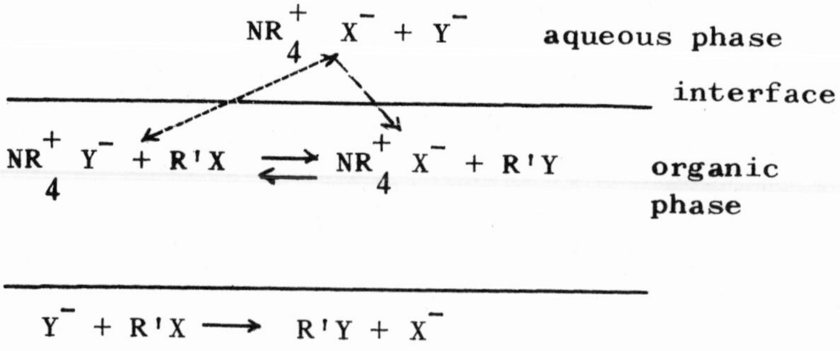

$$NR_4^+ X^- + Y^- \qquad \text{aqueous phase}$$

interface

$$NR_4^+ Y^- + R'X \rightleftharpoons NR_4^+ X^- + R'Y \qquad \text{organic phase}$$

$$Y^- + R'X \longrightarrow R'Y + X^-$$

Fig. 2.   Course of phase-transfer nucleophilic substitution.

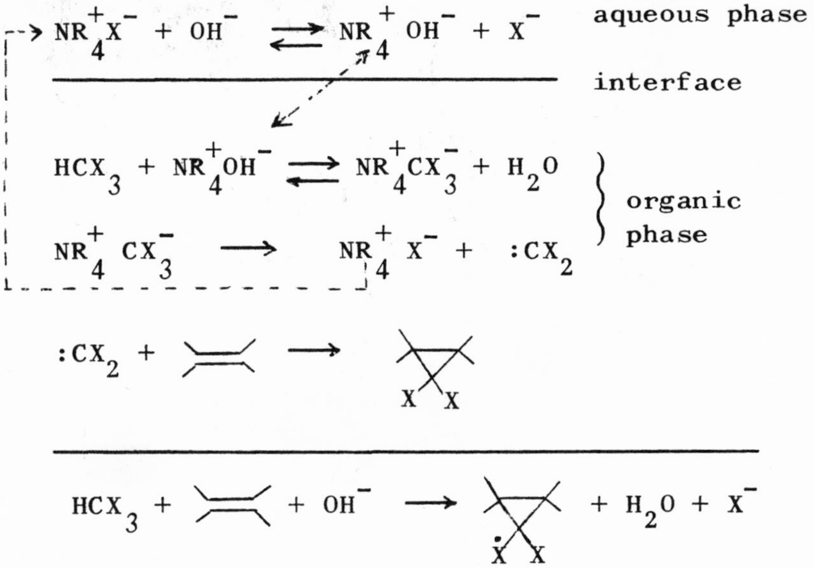

$$NR_4^+ X^- + OH^- \rightleftharpoons NR_4^+ OH^- + X^- \quad \text{aqueous phase}$$

——————————————————————— interface

$$HCX_3 + NR_4^+ OH^- \rightleftharpoons NR_4^+ CX_3^- + H_2O \quad \Big\} \quad \text{organic}$$

$$NR_4^+ CX_3^- \longrightarrow NR_4^+ X^- + \;:CX_2 \quad \Big) \quad \text{phase}$$

$$:CX_2 + \;\rangle\!=\!\langle \longrightarrow \;\rangle\!\!\!\!\times\!\!\!\!\langle$$
$$X \quad X$$

——————————————————————————————————————

$$HCX_3 + \;\rangle\!=\!\langle + OH^- \longrightarrow \;\rangle\!\!\!\!\times\!\!\!\!\langle + H_2O + X^-$$
$$X \quad X$$

Fig. 3.   Course of phase-transfer cyclopropene formation.

Phase-Transfer Technique in Transition Metal Chemistry

So far only very few examples of the use of this technique in homogeneous catalysis and organometallic chemistry are known, but we expect a considerable development of the use of this technique.

The possibility of using the phase transfer technique in homogeneous catalysis with transition metal complexes has been recently shown to have some interesting aspects.

1. Base catalyzed activation of substrates. It has recently been shown that monosubstituted acetylenes are converted into disubstituted acetylenes by reaction with aryl, heterocyclic or vinylic chlorides, bromides and iodides at 50-100°C in the presence of a base such as sodium methoxide or phenoxide or a basic amine and palladium phosphine complexes.[6, 7]

$$ArX + R-C \equiv CH + B \xrightarrow[50-100^\circ C]{Pd[P(C_6H_5)_3]_4} ArC \equiv C-R + B \cdot HX$$

The formation of acetylenic compounds occurs easily at room temperature when aryl palladium(II) or aryl nickel(II) complexes are reacted with the sodium salts of the acetylenic compounds.

$$\underset{\underset{P(C_6H_5)_3}{|}}{\overset{\overset{P(C_6H_5)_3}{|}}{Ar-M-X}} + R-C\equiv C-Na \xrightarrow{\quad P(C_6H_5)_3 \quad} Ar-C\equiv C-R + NaX + M[P(C_6H_5)_3]_3$$

M = Pd, Ni.

This and other evidence suggest the following mechanism for the palladium catalyzed synthesis of acetylenic compounds.

$$Ar-X + Pd[P(C_6H_5)_3]_4 \rightleftharpoons \underset{\underset{P(C_6H_5)_3}{|}}{\overset{\overset{P(C_6H_5)_3}{|}}{Ar-Pd-X}} + 2P(C_6H_5)_3$$

$$R-C\equiv C-H + NaOCH_3 \rightleftharpoons R-C\equiv C^-Na^+ + CH_3OH$$

$$\underset{\underset{P(C_6H_5)_3}{|}}{\overset{\overset{P(C_6H_5)_3}{|}}{Ar-Pd-X}} + R-C\equiv C^-Na^+ \longrightarrow \left[\underset{\underset{P(C_6H_5)_3}{|}}{\overset{\overset{P(C_6H_5)_3}{|}}{Ar-Pd}}\underset{C\equiv C-R}{\overset{X}{<}} \right] Na^+$$

$$\downarrow$$

$$\underset{\underset{Pd[P(C_6H_5)_3]_2}{|}}{Ar-C\equiv C-R} + X^- + Na^+$$

$$\underset{\underset{Pd[P(C_6H_5)_3]_2}{\diagdown}}{Ar-C\equiv C-R} + P(C_6H_5)_3 \rightleftharpoons Ar-C\equiv C-R + Pd[P(C_6H_5)_3]_3$$

Now since the generation of basic condition can be obtained by the phase transfer technique, this reaction has been successfully carried out in a two phase system (sodium hydroxide + benzene) and an alkylammonium salt[8]

$$ArX + R-C\equiv CH \xrightarrow[Pd[P(C_6H_5)_3]_4, NR_4X]{\overset{NaOH}{\underset{}{aqueous}/Benzene}} Ar-C\equiv C-R + NaX$$

The reaction occurs very easily and disubstituted acetylenic compounds are obtained in high yields.

The mechanism of this process probably involves formation of the carbanionic species by the action of alkylammonium hydroxide in the interphase and substitution reaction of this species on the aryl palladium complex formed in the organic phase.

$$\text{(C}_6\text{H}_5\text{)-Br} + \text{(C}_6\text{H}_5\text{)-C}\equiv\text{CH} + \text{NaOH} \xrightarrow[\text{Pd[P(C}_6\text{H}_5)_3]_4,\ \text{NEt}_3\text{benzyl Cl}]{\substack{\text{NaOH}\\ \text{Aqueous}\ /\text{Benzene}}}$$

$$\xrightarrow{\hspace{2cm}} \text{C}_6\text{H}_5\text{-C}\equiv\text{C-C}_6\text{H}_5 \quad + \text{NaBr}$$

Yields of about 90% and about 800 moles of diphenyl acetylene x mole of Palladium complex have been obtained.

Similar results have been obtained with several aromatic and vinylic halides.

This example shows the possibility of using phase-transfer conditions to activate one of the reactants which enters into the catalytic cycle.

2. Carbonylation of organic halides. Another catalytic reaction which can be conveniently carried out using a two phase system is the carbonylation of aromatic and benzylic halides.

The carbonylation of aryl, vinyl and benzyl halides has been shown to occur with several transition metal catalysts.[9]

Recently this reaction has been shown to occur with palladium triphenylphosphine complexes as catalyst in the presence of a base under mild conditions.[10]

$$\text{R-X} + \text{CO} + \text{R'OH} + \text{B} \xrightarrow{\text{Pd complex}} \text{RCOOR'} + \text{B·HX}$$

We have found that this reaction can be conveniently carried out using the phase transfer technique[11]

$$\text{RX} + \text{CO} + \text{NaOH} \xrightarrow[\text{Pd[P(C}_6\text{H}_5)_3]_4,\ \text{NR'}_4\text{X}]{\substack{\text{NaOH}\\ \text{aqueous}\ /\text{Benzene}}} \text{R-COONa} + \text{NaX}$$

The reaction occurs by mixing an aqueous solution of NaOH with a toluene solution of the catalyst in the presence of an alkyl ammonium salt.

The organic halide is added slowly to the reaction mixture under carbon monoxide to give the sodium salt of the organic acid and sodium halide.

In the case of benzyl chloride this process affords phenylacetic acid in high yield.

$$\text{C}_6\text{H}_5\text{-CH}_2\text{Cl} + CO + 2NaOH \xrightarrow[\text{PdCl}_2[\text{P(C}_6\text{H}_5)_3]_3/\text{C}_6\text{H}_5\text{CH}_2\text{N(C}_2\text{H}_5)_3\text{Cl}]{\text{NaOH}_\text{aqeous}/\text{p-xylene}}$$

$$\xrightarrow{\quad\quad} \text{C}_6\text{H}_5\text{-CH}_2\text{COONa} + NaCl$$

Yield as high as 80-90% can be obtained the secondary product being benzyl alcohol.

Very high catalyst turnover can be obtained by using this technique (4.000 moles of phenylacetic acid salt per mole of palladium).

The xylene solution of the palladium complex can be easily recovered after separation of the aqueous phase containing the phenylacetic salts and recycled several times after addition of small quantity of triphenylphosphine.

Under similar conditions the carbonylation of aromatic halides can be carried out.

Sodium benzoate can be easily obtained from bromobenzene.

It is interesting to point out that this technique is very effective for the selective carbonylation of polyhalogenated aromatic halides.

For ezample it has been possible to obtain the sodium salt of p-bromobenzoic acid from dibromobenzene with selectivity as high as 95%.

$$\text{Br-C}_6\text{H}_4\text{-Br} + CO + NaOH \longrightarrow \text{Br-C}_6\text{H}_4\text{-COONa} + NaBr$$

Moreover from 1,3,5-trichlorobenzene it has been possible to obtain the 3,5-dichlorobenzoic acid with selectivity as high as 97%.

This very high selectivity is possible because conversion of the first C-X group into C-COO- causes the molecule to go from the organic to the aqueous phase, thus removing it from further reaction.

Figure 4 shows the probable reactions involved in carbonylation of organic halides (RX), following the general scheme applicable to phase-transfer catalysis.

The nucleophile which attacks the acylpalladium complex is probably X^-, more easily transferred into the organic phase as an ammonium salt.

An important practical aspect of this process is the continuous separation of the products from the catalyst, which in effect heterogenizes the homogeneous catalyst.

This point accounts for the high catalyst turnover, the selectivity encountered in the carbonylation of polyhalogenated compounds, and the high activity of the catalyst.

The carbonylation of organic halides can be achieved under similar phase transfer condition by using $Na[Co(CO)_4]$ as a catalyst.[12, 13]

The cobalt carbonyl anion is transferred into the organic phase as the ammonium salt, while the organic acid goes from the organic to the aqueous phase.

This reaction is however limited to aliphatic organic halide and in particular is very effective for the carbonylation of benzyl halides.

3. Synthesis of organometallic compounds. The application of phase transfer catalysis to the synthesis of organometallic complexes has been recently reported.

The first example is related to the synthesis of iron complexes

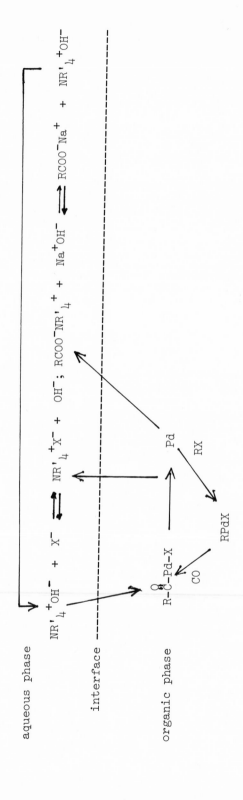

FIG. 4 – Phase transfer catalysis in palladium-catalyzed carbonylation of organic halides. Pd denotes palladium with its associated phosphine.

via ortho-metallation reaction of diiron enneacarbonyl with thio-
benzophenones. This reaction was known to occur in benzene at room
temperature for 30-36 h.[14]

A dramatic enhancement of reaction rate, as well as simpler
reaction work-up, occurred by the use of phase transfer catalysis.[15]
The same reaction can be infact carried out in 30-45 minutes by treatment of a benzene solution of a thiobenzophenone with triirondode-
cacarbonyl, aqueous sodium hydroxide and benzyltriethylammonium
chloride.

The ortho-methalated complexes I are obtained with yields of
36-80% depending on the nature of the R group. Less than 3% ortho-
methallation occurred in the absence of the catalyst.

Phase transfer conditions are likely to induce the formation
of anionic complexes such as $HFe_3(CO)_{11}^-$ which are important species
in the reaction.

The use of the phase transfer catalysis to prepare π-allyl cobalt tricarbonyl complexes has been reported.[16] These complexes are
obtained in 70-80% yields by reaction of a benzene solution of an
allyl bromide with dicobalt octacarbonyl, 5 N NaOH and benzyltriethylammonium chloride as the catalyst.

The phase transfer catalyzed preparation of II is superior to
other syntheses of these complexes in terms of yield, speed, mildness
and simplicity.

The initial step in the reaction is likely conversion of the $Co_2(CO)_8$

to cobalt tetracarbonyl anion in the basic phase transfer condition.

The anionic cobalt species reacts with the allyl halide to give the product.

Since the $Co(CO)_4^-$ ion is a probable intermediate in the formation of alkylidynetricobalt nonacarbonyls from di- and tri-halomethyl compounds and dicobalt actacarbonyl, this reaction has been shown to be ideally suited to phase transfer catalysis techniques.[16]

Treatment of trihalomethyl compound III with $Co_2(CO)_8$, using benzyltriethylammonium chloride as the catalyst, and 3-5 N NaOH in benzene for 0.75-2.5 h at room temperature affords the cluster IV in good yield.

$$R-CX \; + \; Co_2(CO)_8 \quad \xrightarrow[\text{3-5 N NaOH} \quad C_6H_6]{\text{PhCH}_2\text{N(C}_2\text{H}_5)_3\text{Cl}} \quad (CO_3)Co\overset{\overset{\displaystyle R}{\overset{|}{C}}}{+}Co(CO)_3$$

III

$$\underset{Co(CO)_3}{}$$

IV

Conclusions

The application of phase transfer catalysis to organometallic chemistry and homogeneous catalysis is at a very preliminary stage. However the examples already known and here reviewed have indicated some interesting features which can make this technique of great value in transition metal chemistry.

So far the main recognized functions of this technique are:

1. Activate reagents which enter in a catalytic cycle.
2. Remove continuously compounds formed in a catalytic process from the phase containing the catalyst.
3. Activate the organometallic complex by transforming it to more reactive species.
4. Make easier reactions which occur in basic media.

References

1.a. M. Makosza, B. Serafin, T. Urbanski, Chimie et Industrie (Paris), 93, 537 (1965);
  b. C. M. Starks, J. Am. Chem. Soc., 93, 195 (1971);
  c. M. Makosza, Pure Appl. Chem., 43, 439 (1975);
  d. J. Dockx, Synthesis, 1973, 441;
  e. E. V. Delmon, Angew. Chem. Internat. Ed. Engl., 13, 170 (1974); Chem. Techn., 1975, 210.

2. D. T. Sam, H. E. Simmons, J. Am. Chem. Soc., 96, 2252 (1974);
C. L. Liotta, H. P. Harris, ibid., 96, 2250 (1974); M. Makosza,
M. Ludwikow, Angew. Chem., 86, 744 (1974); D. Landini, F. Montanari,
Chem. Comm., 1974, 879; M. Cinquini, F. Montanari, P. Tundo, ibid.,
1975, 393.

3. A. J. Parker, Chem. Rev., 69, 1 (1969).

4. M. Makosza, B. Serafinova, Rocz. Chem., 39, 1223, 1401, 1595,
1799, 1805 (1965); 40, 1647, 1839 (1966).

5. M. Makosza, W. Wawrxyniewicz, Tetrahedron Lett., 1969, 4659; E.
V. Delmon, J. Schönefeld, Liebigs Ann. Chem., 744, 42 (1971).

6. L. Cassar, J. Organometal. Chem., 93, 253 (1975).

7. R. Heck, J. Organometal. Chem., 93, 258 (1975).

8. L. Cassar, in preparation.

9. T. Weil, L. Cassar, M. Foà, in "Organic Syntheses via Metal
Carbonyls," Vol. II, I. Wender and P. Pino, ed., Interscience, New
York, to be published.

10. A. Schoenberg, I. Bartoletti and R. F. Heck, J. Org. Chem., 39,
3318 (1974); J. K. Stille, L. F. Hines, R. W. Friess, P. K. Wong,
D. E. Tanees and K. Lau in Advances in Chemistry Series No. 132,
Amer. Chem. Soc., 90, 1974; M. Hidai and Y. Uchida in Y. Ishii
and M. Tsutsui, Organotransition-Metal Chemistry, Plenum Press,
New York-London 1975, p. 265; J. K. Stille, P. W. Wong, J. Org.
Chem., 40, 532 (1975); M. Hidai, T. Hikita, Y. Wada, Y. Fujikura
and Y. Uchida, Bull. Chem. Soc. Japan, 48, 2075 (1975).

11. L. Cassar, M. Foà, A. Gardano, J. Organometal. Chem., 121, C55
(1976).

12. L. Cassar, M. Foà, in preparation.

13. H. Alper, private communication.

14. H. Alper, A. S. K. Chan, J. Amer. Chem. Soc., 97, 4251 (1973).

15. H. Alper, D. Des Roches, J. Organometal. Chem. 117, C44 (1976).

16. H. Alper, H. Des Abbayes, D. Des Roches, J. Organometal. Chem.,
121, C31 (1976).

THE ACTIVATION OF MOLECULAR NITROGEN

John E. Bercaw

Division of Chemistry and Chemical Engineering
California Institute of Technology
Pasadena, California 91125

Introduction

In view of its abundance and availability molecular nitrogen is potentially a valuable substrate for catalytic reactions. Its present limited usage in industry is as a feedstock in the Haber-Bosch ammonia synthesis, a process requiring a heterageneous catalyst, high temperatures, and high pressures. The selective reduction of N_2 to hydrazine or to organic nitrogen compounds would be highly desirable but may well be possible only with homogeneous catalysts operating under mild conditions. Thus the development of transition metal compounds capable of promoting the reduction of N_2 remains a challenge to inorganic and organometallic chemists.

Following their discovery in 1965, a number of transition metal dinitrogen complexes have been prepared; however, in the great majority of these complexes the coordinated N_2 has resisted all attempts at reduction. In other soluble transition metal based systems the reduction of N_2 to ammonia or hydrazine has been achieved, yet the nature of the reaction intermediates, particularly transient dinitrogen complexes, remains a mystery for the most part. During the past decade research in abiological nitrogen fixation has proceeded along two lines: (1) the search for highly efficient, mechanistically more transparent N_2 reducing systems, and (2) the synthesis, isolation, and characterization of transition metal dinitrogen complexes which contain N_2 ligands activated toward subsequent chemical reduction. The common goal is the definition of the chemistry available to coordinated dinitrogen, a logical prerequisite for the rational design of homogeneous catalytic N_2 reducing systems.

129

Homogeneous N_2 Reducing Systems

The presently known transition metal based N_2 reducing systems may be divided into two catagories: (1) those in which the metal dinitrogen complex intermediate have not yet identified, and (2) those which begin with a well characterized dinitrogen complex.

Among the earliest N_2 reducing systems, representative of the former, are mixtures of early transition halides or alkoxides and strong reducing agents in aprotic media, e.g. $MCl_3(M=V,Cr,Fe)/$ $Li^+C_{10}H_8^-$, $Ti(OR)_2/Na^+C_{10}H_8^-$.[1-14] Such systems efficiently reduce N_2 at room temperature apparently yielding stable metal nitrides, which may be subsequently hydrolyzed to ammonia. Furthermore, it has been shown that treatment of the product of Mg reduction of $(C_5H_5)_2TiCl_2$ under N_2 (presumably a titanium nitride) with aldehydes or ketones leads to primary or secondary amines in yields approaching 50%.[15]

Shilov and coworkers have reported the reduction of N_2 to hydrazine or ammonia in protic media.[16-21] One such system utilizes $MoOCl_3$ or MoO_4^{2-} as the catalyst precursor for the reduction of N_2 by Ti^{3+}, Cr^{2+}, or V^{2+} hydroxides.[16,19] While the reaction is catalytic with respect to molybdenum, these systems are heterogeneous and the reaction rates are strongly dependent on the presence of magnesium, calcium, or barium ions in the hydroxide precipitate serving as catalyst. Later it was found that V^{2+}/Mg^{2+} hydroxide gels efficiently reduce N_2 to N_2H_4 in protic media at room temperature and atmospheric pressure, even in the absence of molybdenum sales.[17,18]

Most recently Shilov has reported a soluble $V^{2+}/catechol$ complex which reduces N_2 to ammonia in methanol or water-methanol mixtures at 25°.[20,21] The stoichiometry of the reaction at sufficiently high N_2 pressures approaches

$$N_2 + 8V^{2+} + 8H_2O \longrightarrow 2NH_3 + H_2 + 8V^{3+} + 8OH^- \qquad (1)$$

The kinetics of this reaction were interpreted as involving an $8e^-$ transfer (or two coupled $4e^-$ transfers) with six used in the reduction of N_2 to NH_3 and the remaining two for the reduction of H_2O to H_2. This result and other evidence which indicate aggregation of the vanadium catechol complex in solution suggest that four of eight V(II) ions are present in the transition state of the reaction.

Parallel with these developments it was demonstrated that ligated N_2 in two types of dinitrogen complexes, (1) $ML_4(N_2)_2$ (M=Mo, W; L=PR$_3$) and (2) $\{(C_5Me_5)_2MN_2\}_2N_2$ (M=Ti,Zr), may be protonated and reduced to ammonia and hydrazine. In that these two types are the first well characterized dinitrogen complexes capable of liberating reduced N_2 on simple protonation, the investigation of their molecular and electronic structures and of the mechanisms of N_2 reduction could point the way to homogeneous catalytic processes utilizing N_2 as a substrate.

Molybdenum and Tungsten Dinitrogen Complexes

Chatt and coworkers obtained their first indication that the dinitrogen ligands in complexes of the type $M(PR_3)_4(N_2)_2$ ($M=Mo$,W) might be reactive from experiments with acetyl chloride.[22] The reaction proceeds according to equation 2 (dpe= $Ph_2PCH_2CH_2PPh_2$):

$$\text{trans-}[W(dpe)_2(N_2)_2] + CH_3COCl + HCl$$
$$\searrow \qquad\qquad\qquad\qquad (2)$$
$$[W(dpe)_2Cl_2\{N_2H(COCH_3)\}] + N_2$$

Immediately following this finding it was discovered that the N_2 could be protonated directly with HCl or better with HBr.[23]

$$\text{trans-}[W(dpe)_2(N_2)_2] + 2HBr$$
$$\searrow \qquad\qquad\qquad\qquad (3)$$
$$[W(dpe)_2Br_2(N_2H_2)] + N_2$$

The product of this reaction contains one labile Br^-, which may be displaced by tetraphenylborate to yield a crystalline cationic complex.

$$[W(dpe)_2Br_2(N_2H_2)] + BPh_4^-$$
$$\searrow \qquad\qquad\qquad\qquad (4)$$
$$\text{trans-}[W(dpe)_2Br(NNH_2)]^+BPh_4^- + Br^-$$

The x-ray structure determination of this compound indicated a linear WNN arrangement and a short WN distance, thus suggesting the following structure:

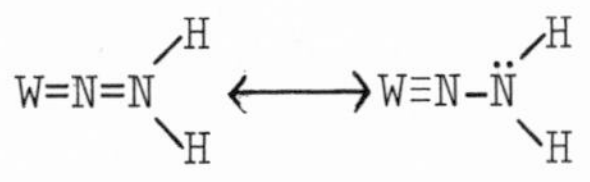

$$W=N=N\overset{\diagup H}{\diagdown H} \longleftrightarrow W\equiv N-\overset{\cdot\cdot}{N}\overset{\diagup H}{\diagdown H}$$

A diazene substructure is adopted in the 7-coordinate neutral complex.[25-27]

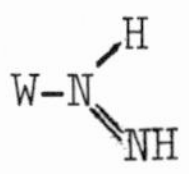

$$W-N\overset{\diagup H}{\diagdown NH}$$

All attempts by the Chatt group to further reduce these N_2H_2 complexes to ammonia or hydrazine failed, despite the great variety of reagents used.[25] However, when an analogous complex containing monodentate phosphines (e.g. $\underline{cis}$-$[W(PMe_2Ph)_4(N_2)_2]$) was treated with sulfuric acid in methanol, ammonia was obtained in yields equal to[25,28] 90% of that expected according to equation 5.

$$\text{cis-}\left[W(PMe_2Ph)_4(N_2)_2\right] \xrightarrow[\text{MeOH}]{H_2SO_4} 2NH_3 + N_2 + \text{"W(VI) products}$$
$$(5)$$

Lower yields were obtained for cis-$\left|Mo(PMePh_2)_4(N_2)_2\right|$, trans-$\left[W(PMePh_2)_4(N_2)_2\right]$, and trans-$\left[Mo(PMePh_2)_4(N_2)_2\right]$. In aprotic media (e.g. THF) considerable amounts of hydrazine were evloved along with ammonia.

Immediately following these reports from the Sussex group Brulet and Van Tamelen reported the successful protonation and reduction of the N_2 to ammonia (0.37 mol/mol Mo) in trans-$\left[Mo(dpe)_2(N_2)_2\right]$.[29] The key features of their system are apparently the highly polar reaction media (N-methylpyrrolidone or propylene carbonate) in combination with HBr. The species $Mo(dpe)_2Br_2(N_2H_2)$ was reported to yield 0.15 mol NH_3/mol Mo under the same conditions. That a mixture of $Mo(dpe)_2Br_2(N_2H_2)$ and trans-$\left[Mo(dpe)_2(N_2)_2\right]$ gave a significantly higher yield of NH_3 (0.48 mol/mol total Mo) was taken to suggest that a second molybdenum serves as a reducing agent or that the reduction is mediated by a binuclear complex.[29]

More recently the Sussex group reported the alkylation of one ligated N_2 in trans-$\left[M(dpe)_2(N_2)_2\right]$ (M=Mo,W) when benzene solutions are irradiated in the presence of alkyl halides (eq. 6):[25,30,31]

$$\text{trans-}\left[W(dpe)_2(N_2)_2\right] + RBr \xrightarrow{h\nu} \text{trans-}\left[W(dpe)_2Br(N_2R)\right] + N_2$$
$$(6)$$

$$(R=CH_3, C_2H_5, C_6H_{11})$$

Furthermore, under certain conditions the distal N atom of the ligated N_2 may be alkylated and protonated ao even dialkyated, for example with excess methyl iodide or $Br(CH_2)_4Br$ (eqs. 7 and 8):[25]

$$\text{trans-}\left[W(dpe)_2Br(N_2R)\right] + HBr \searrow \left[W(dpe)_2Br(N_2HR)\right]^+Br^-$$
$$(7)$$

$$\text{trans-}\left[W(dpe)_2(N_2)_2\right] + \overset{Br}{\underset{Br}{\diagdown\diagup}} \xrightarrow{h\nu} \left[(dpe)_2BrW=N-N\bigcirc\right]^+Br^-$$
$$(8)$$

Titanium and Zirconium Dinitrogen Complexes

The relatively high efficiency of titanocene-based N_2 reducing systems has stimulated much research since their discovery over a decade ago.[32] It has been demonstrated that titanocene, $\{(C_5H_5)_2Ti\}_n$ (n=1 or 2), is an intermediate common to many of these reaction mixtures, and that this complex absorbs N_2 in a rapid, reversible reaction to yield the blue binuclear dinitrogen complex

$\{(C_5H_5)_2Ti\}_2N_2$.[6,7,33-35] Other investigators have found that Ti(III) complexes of the type $(C_5H_5)_2TiR$ (R=C_6H_5,$CH_2C_6H_5$,i-C_3H_7) also reversibly absorb N_2 to form diamagnetic, blue binuclear dinitrogen complexes $\{(C_5H_5)_2TiR\}_2N_2$.[36-39] Recently Pez and Kwan reported that $\{(C_5H_5)_3(C_5H_4)Ti_2\}$ reversibly binds dinitrogen to yield again a blue complex formulated as $\{(C_5H_5)_3(C_5H_4)Ti_2\}_2N_2$.[40,41] All three types of dinitrogen complexes liberate ammonia in good yield when treated first with naphthalide, than with HCl. Furthermore, under appropriate conditions $\{(C_5H_5)_2Ti\}_2N_2$ may be protonated directly to yield either NH_3 and N_2 or N_2H_4 and N_2.

It has not yet been possible, however, to isolate a crystalline dinitrogen complex of titanocene suitable for definitive structural characterization. The nature of these "active" dinitrogen complexes and their relationship to other structurally characterized, but inert, dinitrogen complexes have thus been subjects of considerable interest and speculation. The difficulties encountered when attempting to isolate stable crystalline forms of $\{(C_5H_5)_2Ti\}_2N_2$ and $\{(C_5H_5)_2TiR\}_2N_2$ may be traced, at least in part, to facile cyclopentadienyl-to-titanium hydrogen transfer inherent to these systems. We have therefore investigated the closely related pentamethylcyclopentadienly derivatives. We find compounds containing the $(\eta^5$-$C_5Me_5)_2M$ (M=Ti,Zr) moiety often more stable and much more amenable to study than their $(\eta^5$-$C_5H_5)$ analogs.

Synthesis and Characteetization of Dinitrogen Complexes of $(\eta^5$-$C_5Me_5)_2Ti$ and $(\eta^5$-$C_5Me_5)_2Zr$. The starting materials, $(\eta^5$-$C_5Me_5)_2$ $TiCl_2$ and $(\eta^5$-$C_5Me_5)_2ZrCl_2$, were prepared in a similar manner according to equations (9) and (10).

$$2\ Li(C_5Me_5) + TiCl_3 \xrightarrow{\hspace{1cm}} \quad HCl \hspace{4cm} (9)$$

$$(\eta^5\text{-}C_5Me_5)_2TiCl_2 + 2\ LiCl + \tfrac{1}{2}H_2$$

$$2\ Li(C_5Me_5) + ZrCl_4 \longrightarrow (\eta^5\text{-}C_5Me_5)_2ZrCl_2 + 2\ LiCl \quad (10)$$

Since direct reduction of $(\eta^5$-$C_5Me_5)_2TiCl_2$ proceeded at a rate comparable to that for the decomposition of the reduced products, an indirect route,[44] equations (11)-(14), was employed to generate $(\eta^5$-$C_5Me_5)_2Ti$.

$$(\eta^5\text{-}C_5Me_5)_2TiCl_2 + 2\ LiCH_3 \xrightarrow{\hspace{2cm}} \hspace{3cm} (11)$$

$$(\eta^5\text{-}C_5Me_5)_2Ti(CH_3)_2 + 2\ LiCl$$

$$(\eta^5\text{-}C_5Me_5)_2Ti(CH_3)_2 \xrightarrow{\quad \Delta \quad} \hspace{3cm} (12)$$

$$(\eta^5\text{-}C_5Me_5)(C_5Me_4CH_2)TiCH_3 + CH_4$$

$$(\eta^5\text{-}C_5Me_5)(C_5Me_4CH_2)TiCH_3 + 2H_2 \searrow$$

$$(\eta^5\text{-}C_5Me_5)_2TiH_2 + CH_4 \qquad (13)$$

$$(\eta^5\text{-}C_5Me_5)_2TiH_2 \rightleftharpoons (\eta^5\text{-}C_5Me_5)_2Ti + H_2 \qquad (14)$$

Exposure of $(\eta^5\text{-}C_5Me_5)_2Ti$ to N_2 in pentane at $0°$ yields crystalline $\{(\eta^5\text{-}C_5Me_5)_2Ti\}_2N_2$ (1). 1 is a dark blue complex

$$2(\eta^5\text{-}C_5Me_5)_2Ti + N_2 \rightleftharpoons \{(\eta^5\text{-}C_5Me_5)_2Ti\}_2N_2 \qquad (15)$$

(λ_{max}=512, 642 nm) and exhibits no adsorption in the infrared attributable to an NN stretching mode. 1 is paramagnetic (μ_{eff}=2.18 (1) BM per Ti), and a single resonance is observed in the ^1H nmr spectrum which is paramagnetically shifted 60.9 ppm downfield of TMS at $-55°$.

The structure of 1 as determined by single crystal x-ray diffraction methods[46] is shown in Figure 1. As can be seen the structure consists of two staggered $(\eta^5\text{-}C_5Me_5)_2Ti$ units bridged by the N_2 in a linear Ti-N≡N-Ti arrangement. N-N distances for the two molecules in the asymmetric unit (space group P1̄), are 1.165(14) and 1.155(14) Å, somewhat longer than those for other binuclear dinitrogen complexes.[47] The conventional end-on N_2 bonding mode is of special significance in view of earlier suggestions that either an edge-on[34,43,48] or doubly bent (azo)[33] coordination should be favored for complexes of this type.

On cooling toluene solutions of $(\eta^5\text{-}C_5Me_5)_2Ti$ or 1 below ca. $-20°$ under 1 atm, the purple-blue (λ_{max}=578 nm) complex $\{(\eta^5\text{-}C_5Me_5)_2TiN_2\}_2N_2$ (2) is formed. Although 2 is not sufficiently stable for its isolation as a stable crystalline material, it has been studied in situ below $-50°$. ^1H and ^{15}N nmr and ir spectra for 1 are indicative of a structure identical to that for $\{(\eta^5\text{-}C_5Me_5)_2ZrN_2\}_2N_2$ (see below).

Reduction of $(\eta^5\text{-}C_5Me_5)_2ZrCl_2$ with sodium amalgam in toluene under N_2 leads to a 50-60% yield of $\{(\eta^5\text{-}C_5Me_5)_2ZrN_2\}_2N_2$.[44]

$$2(\eta^5\text{-}C_5Me_5)_2ZrCl_2 + 4Na \searrow N_2$$

$$\{(\eta^5\text{-}C_5Me_5)_2ZrN_2\}_2N_2 + 4 NaCl \qquad (16)$$

The structure of 3 as determined by single-crystal x-ray diffraction methods[49] is illustrated in Figure 2. As can be seen the binuclear structure consists of two $(\eta^5\text{-}C_5Me_5)_2ZrN≡N$ moieties bridged by a third dinitrogen ligand. Terminal and bridging dinitrogen ligand are bonded end-on in essentially linear Zr-N≡N and Zr-N≡N-Zr arrangements. The two terminal N-N bond lengths (1.116(8) and 1.114(7) Å) are within the range of those found

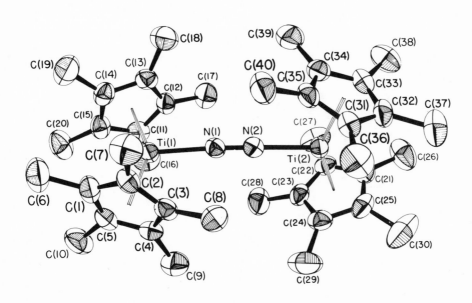

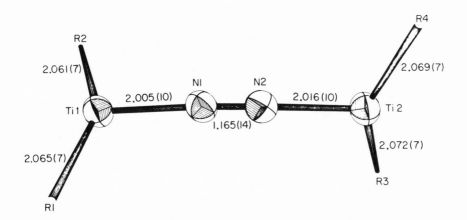

Figure 1.   The molecular structure of
$\{(\eta^5\text{-}C_5Me_5)_2Ti\}_2N_2$.

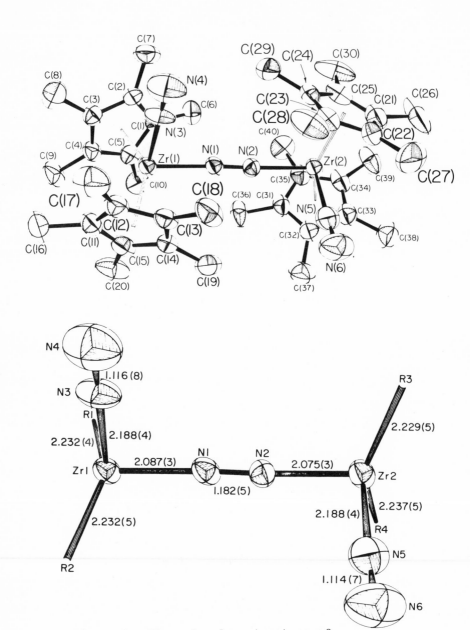

Figure 2.   The molecular structure of
$\{(\eta^5\text{-}C_5Me_5)_2ZrN_2\}_2N_2$.

previously for mononuclear dinitrogen complexes $(1.11 \pm 0.01$ Å$)$,[45] only slightly expanded from that for free N_2 $(1.976$ Å$)$. The bridging dinitrogen ligand, however, exhibits a significantly longer N-N distance $(1.182$ (5) Å$)$, indicative of a substantial reduction in bond order.

The infrared spectrum of 3 (Nujol mull) exhibits two strong bands at 2041 and 2006 cm^{-1} and a band of medium intensity at 1556 cm^{-1}, which shift upon substitution of doubly labeled $^{15}N_2$ to 1972, 1937, and 1515 cm^{-1}, respectively. This spectrum is interpreted on the basis of a coupling of the symmetric combination of terminal N-N stretches $(2041$ $cm^{-1})$ with the bridging N-N mode $(1556$ $cm^{-1})$, the latter gaining intensity fron tha former, the strongest band $(2006$ $cm^{-1})$ is attributed to the anti-symmetric combination of terminal N-N modes. The observation of two terminal N-N stretching frequencies is significant in that it indicates of substantial electronic coupling of the two $(\eta^5-C_5Me_5)_2ZrN\equiv N$ units via the $\mu-N_2$ (the nearly orthogonal spatial relationship of the terminal N_2 ligands eliminates the possibility of through-space dipolar coupling).

The ^1H nmr spectrum of 3 at $5°$ (toluene-d_8) shown the expected two singlets attributable to the pairwise equivalent $(\eta^5-C_5Me_5)$ rings. ^{15}N nmr data for $\{(\eta^5-C_5Me_5)_2Zr(^{15}N_2)\}_2(^{15}N_2)$ also support a solution structure for 3 identical with that in the crystalline state. Thus at $-28°$ (toluene-d_8), the 18.25 MHz ^{15}N nmr spectrum for $3-(^{15}N_2)_3$ consists of two doublets attributable to the two ^{15}N nuclei of the two equivalent terminal dinitrogen ligands $(^1J_{^{15}N-^{15}N}= 6.2$ Hz$)$ centered 89.8 and 160.4 ppm upfield of a third singlet resonance due to the two ^{15}N nuclei of the $\mu-N_2$. The ^{15}N nmr spectrum for $3-(^{15}N\equiv^{14}N)_3$ xhibits the same spectrum with the exception that the two upfield doublets now appear as the expected singlets.

Variable temperature ^1H and ^{15}N nmr experiments are in accord with $(\eta^5-C_5Me_5)$ ring site exchange involving step-wise dissociation-accociation of the two terminal dinitrogen ligands of 3. In the temperature range $-26°$ to $+45°$ the two singlets $(1.769$ δ and 1.788 $\delta)$ in the ^1H nmr spectrum steadily broaden, coalesce at $+11°$, and finally give rise to a single line which narrows to $^\sim+45°$ $(1.796_5\delta)$. Over the same temperature range the two doublets due to the ^{15}N nuclei of the two equivalent terminal dinitrogen ligands for $3-(^{15}N_2)_3$ steadily broaden and move slightly upfield toward the chemical shift of free $^{15}N_2$ in toluene-d_8. The signal due to the $\mu-N_2$ remains a sharp singlet over the entire range of temperatures.

These ^1H and ^{15}N nmr data can be fit to the process shown in Figures 3 and 4.[50] Thus (Figure 3) a terminal N_2 dissociates to enerate the intermediate $(\eta^5-C_5Me_5)_2Zr(N_2)-N_2-Zr(\eta^5C_5Me_5)_2$, which may then add N_2 to either side of the trigonal Zr center. A series of two such sequences (Figure 4) may effect complete $(\eta^5-C_5Me_5)$ ring exchange between the two types of sites (α, eclipsed with N_2, and β), and also subjects the ^{15}N nuclei to a weighted average of each of three sites, two bonded to Zr plus that for free N_2 in toluene-d_8.

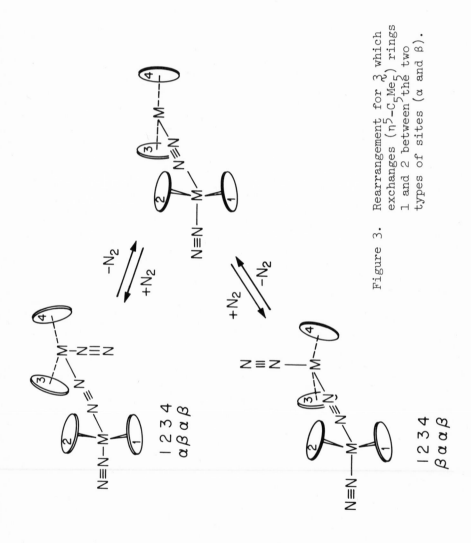

Figure 3. Rearrangement for 3 which exchanges (η^5-C_5Me_5) rings 1 and 2 between the two types of sites (α and β).

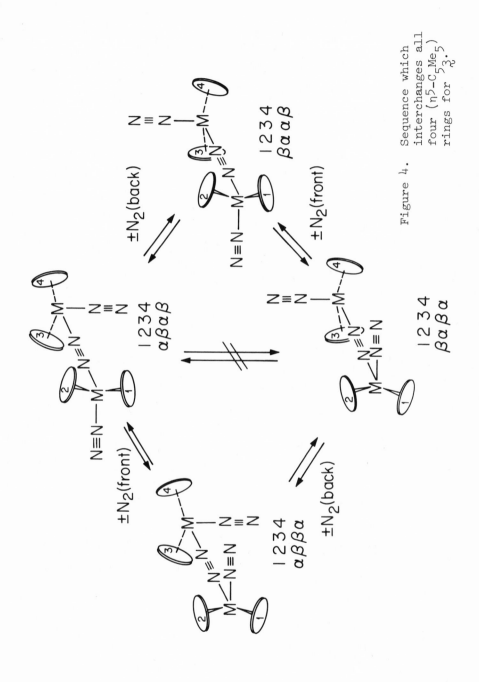

Figure 4. Sequence which interchanges all four $(\eta^5\text{-}C_5Me_5)$ rings for $\underset{\sim}{3}$.

$\underline{N_2\ \text{Reduction to Hydrazine}}$. Treatment of 2 or 3 with anhydrous HCl in toluene at $-80°$ yields $(\eta^5-C_5Me_5)_2TiCl_2$ or $(\eta^5-C_5Me_5)_2ZrCl_2$, respectively, N_2, N_2H_4, and a small amount of H_2 (Table I). The stoichiometry for the reaction with 3 is in close agreement with equation (17) wherein the four reducing equivalents

$$\{(\eta^5-C_5Me_5)_2ZrN_2\}_2N_2 + 4\ HCl$$

$$2(\eta^5-C_5Me_5)_2ZrCl_2 + 2N_2 + N_2H_4 \tag{17}$$

available in the dimer are utilized in the reduction of one of the three N_2 ligands to N_2H_4. Identical treatment of 1 with HCl again leads to hydrazine, but in much lower yield (Table I). This observation presents some intriguing questions concerning the mechanism of hydrazine formation in the reaction of 2 or 3 with HCl... in particular, which of the two types of N_2 (terminal or bridged) is reduced to hydrazine.

The high lability of the terminal dinitrogen ligands for 3 observed in the nmr experiments suggested that it should be possible to prepare labeled specifically with ^{15}N.[51] After 15 minutes exposure of $3-(^{15}N_2)_3$ to free $^{14}N_2$ (toluene, $-23°$), the composition of the gas phase above the solution is in close agreement with that predicted for complete exchange of two of the three dinitrogen ligands of 3. Thus under these conditions exchange occurs between free N_2 and only the two equivalent terminal dinitrogen ligands. incomplete exchange is observed after 5 minutes, permitting an estimate of the terminal N_2 exchange half-life of 2.6(1) minutes ($\tilde{p}_{N_2} = 1$ atm). On the basis of the experiment conducted at reduced pressure the exchange rate appears proportional to free N_2 concentration ($t_{\frac{1}{2}} = 5.0(3)$ minutes, $\tilde{p}_{N_2} = 0.50$ atm).

The results of these exchange experiments allow an accurate assessment of the extent of ^{15}N labeling in both terminal and bridge positions for 3, and thus should allow a determination of which of the three dinitrogen ligands of 3 is reduced to hydrazine in the reaction with HCl. Following exchange of free $^{14}N_2$ with the terminal positions of $3-(^{15}N_2)_3$, a 20 M excess of HCl was added at $-196°$. On warming to $-80°$ an immediate reaction accompanied toluene melting. Evolved N_2 and hydrazine (after IO_3^- oxidation to N_2) were subsequently analyzed by mass spectrometry.

The results clearly indicate that the dinitrogen ligand which is reduced to hydrazine is not exclusively the $\mu-N_2$ as might be anticipated. Rather the composition of both evolved N_2 and hydrazine are in close agreement with those expected for the reaction proceeding according to equation (18).

$$\{(\eta^5-C_5Me_5)_2Zr(N_2^t)\}_2 + 4\ HCl$$

$$2(\eta^5-C_5Me_5)_2ZrCl_2 + 3/2\ N_2^t + 1/2\ N_2^b \tag{18}$$

$$+ 1/2\ N_2^t H_4 + 1/2\ N_2^b H_4$$

Table I. Product distribution for the reactions of $\underset{\sim}{1}$, $\underset{\sim}{2}$, and $\underset{\sim}{3}$ with HCl in toluene at $-80°C$.

	N_2/dimer released	N_2/dimer reduced	N_2H_4/dimer	$2NH_3$/dimer
$\{(\eta^5\text{-}C_5Me_5)_2Ti\}_2N_2$ $(\underset{\sim}{1})$	0.79	0.21	0.15	0.07
$\{(\eta^5\text{-}C_5Me_5)_2TiN_2\}_2N_2$ $(\underset{\sim}{2})$	2.28	0.72	0.66	0.03
$\{(\eta^5\text{-}C_5Me_5)_2ZrN_2\}_2N_2$ $(\underset{\sim}{3})$	2.02	0.98	0.86	0.12

Furthermore, we have found that $\{(\eta^5\text{-}C_5Me_5)_2Zr(CO)\}_2N_2$ $(\underset{\sim}{4})$ may be obtained via treatment of $\underset{\sim}{3}$ with CO at $-20°$. On the basis of its ir spectrum (ν(CO) 1902 (ms), 1860(s), ν(NN) 1682 (ms)), and nmr ((toluene-d_8)s, $\delta 1.80$ (30H)) the structure of $\underset{\sim}{4}$ appears to be identical with $\underset{\sim}{3}$ with CO substituted for the terminal N_2 ligands. Significantly, treatment of $\underset{\sim}{4}$ with HCl as before yields 1 mol of N_2 (no detectable N_2H_4 or NH_3). This observation and the results of the N_2 labeling experiments are indications that the terminal N_2 ligands are playing roles beyond that of mere spectators in the reaction of $\underset{\sim}{3}$ with HCl.

The data require a reaction sequence mediated by a symmetric species in which one terminal N_2 and the μ-N_2 have become equivalent. While a number of mechanisms satisfying this requirement could be formulated, we favor one involving protonation of a terminal dinitrogen of $\underset{\sim}{3}$, loss of the other terminal N_2, and generation of the symmetric reaction intermediate $(\eta\text{-}C_5Me_5)_2Zr(N_2H)_2$ (Scheme I). Consistent with labeling experiments, $(\eta5\text{-}C_5Me_5)_2Zr(N_2H)_2$ would then lead to one mole each of N_2 and N_2H_4. Generation of the neutral, monomeric species $(\eta^5\text{-}C_5Me_5)_2Zr(N_2H)_2$ from $\underset{\sim}{3}$ would require a formal two-electron transfer to the N_2-bearing Zr accompanied by release of the other Zr in the fully oxidized state, i.e., as $(\eta^5\text{-}C_5Me_5)_2$ $ZrCl_2$. Strong electronic coupling of the two Zr(II) centers through the μ-N_2 of $\underset{\sim}{3}$, as indicated by its structure, ir, and visible spectra, should facilitate such Zr-Zr charge transfer.

Whether or not the decomposition of $(\eta^5\text{-}C_5Me_5)_2Zr(N_2H)_2$ to N_2 and N_2H_4 involves free diimide remains an unresolved question. The reaction of $\underset{\sim}{3}$ with HCl was carried out at $-80°$ in the presence of a large excess of azobenzene or dimethylacetylene dicarboxylate in order to trap free N_2H_4, but in neither case was a significant change in stoichiometry observed.[52] However, if the activation energies for the hydrogenation of azobenzene and dicarboxyacetylenes are much greater than that of the disproportionation of diimide (as is likely), no change in stoichiometry would be expected at $-80°$. Unfortunately both PhN=NPh and $MeO_2CC\equiv CCO_2Me$ react directly with $\underset{\sim}{3}$ at higher temperatures where more conclusive results could be obtained.

The formation of ammonia in the reaction of $\underset{\sim}{3}$ with HCl in toluene

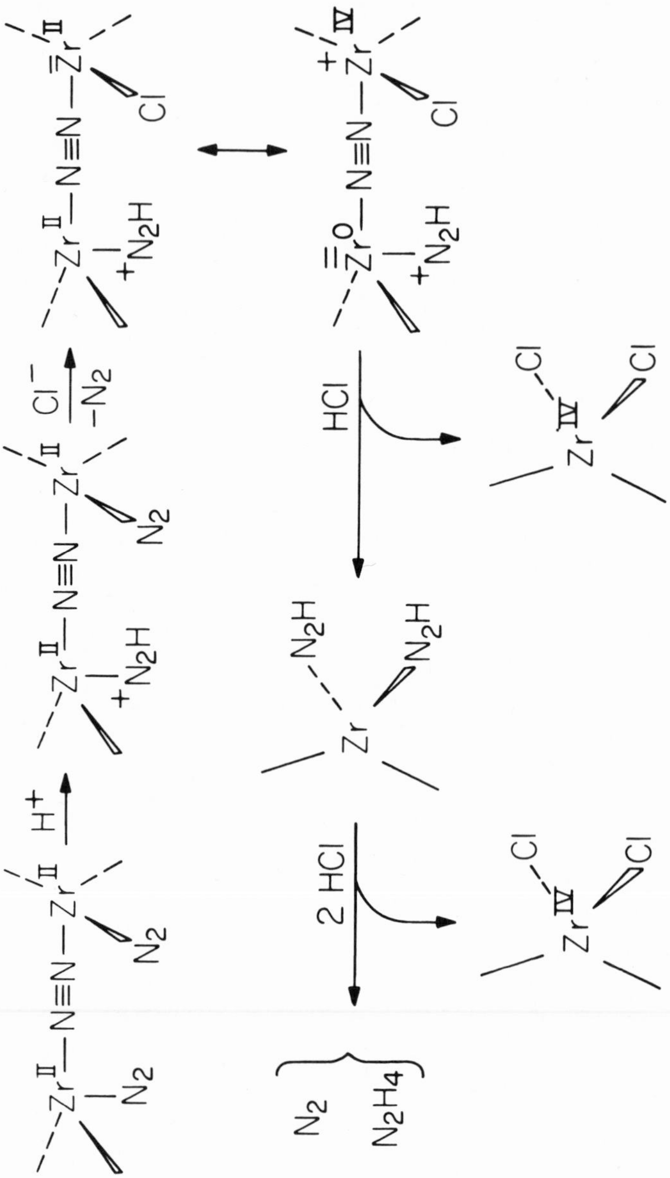

Scheme I

is believed to arise from interference of a reaction of hydrazine
with 3 during the latter stages of the reaction; treatment of $\underset{\sim}{3}$ with
excess anhydrous N_2H_4 at $-80°$ in toluene instantaneously yields N_2
(3 mol) and NH_3(2 mol).

The yields of ammonia and hydrazine from $\underset{\sim}{3}$ depend markedly on
the nature of acid and solvent (Table II). Similar observations
were noted for complexes of the type $ML_4(N_2)_2$ (M=Mo, W; L=PR_3).
There appears to be no obvious interpretation of these effects in
either case; however, it is likely that both the pK_a of the acid
and affinity for the conjugate to bind the metal are important
factors.

Conclusions

These recent advances indicate that we are now begining to
understand some of the essential features of N_2 activation by transition metals. It is interesting to note the similarities between
the two types of N_2 complexes thus far shown to promote N_2 reduction:
(1) $M(PR_3)_4(N_2)_2$ (M=Mo, W) and (2) ${(C_5Me_5)_2MN_2}_2N_2$ (M=Ti, Zr).
Firstly, both types formally contain two dinitrogen ligands bonded
to a single metal atom in the complex. In both cases only one N_2
is reduced, the other is evolved as free N_2 at some point in the reaction sequence. It will be interesting to ascertain whether a
second N is a general requirement for N_2 activation. A possible
exception is ${(C_5Me_5)_2Ti}_2N_2$ which yields small amounts of ammonia
and hydrazine when treated with HCl at $-80°$. It may well be, however,
that the active species in this reaction is, in fact, ${(C_5Me_5)_2TiN_2}_2$
N_2 which is generated from ${(C_5Me_5)_2Ti}_2N_2$ and evolved N_2 during the
stages of the reaction. Secondly, the reduction of N_2 to NH_3 or
N_2H_4 is promoted by electrophiles (HCl, H_2SO_4, RX, RCOX) rather than
by nucleophiles (H$^-$, LiR, etc). The initial step in the N_2 reduction
sequence for both types is apparently electrophilic attack at the
distal N atom of one terminal N_2, with concomitant electron transfer
from transition metal to N_2.

Clearly much more work is required before the reactivity of
ligated N_2 is fully understood. A systematic study of chemistry
available to transition metal dinitrogen complexes, other than reduction to N_2H_4 or NH_3 would be of great value. Such investigations
could ultimately lead to the development of homogeneous catalysts
for the direct conversion of N_2 to organic nitrogen-containing
compounds or for the oxidative fixation of N_2 to nitrate.

Table II. Composition of reduced N_2 (N_2H_4/NH_3) from $\{(\eta^5\text{-}C_5Me_5)_2ZrN_2\}_2N_2$ ($\underset{\sim}{3}$) as a function of acid and solvent.

Solvent	Acid	mol N_2H_4/mol dimer	mol NH_3/mol dimer	% available reducing equivalents as N_2H_4 and NH_3
toluene	HCl	0.86(2)	0.23(4)	92(4)%
	HBr	0.80(2)	0.13(4)	90(4)%
	H_2SO_4	0.46(2)	0.20(4)	61(4)%
diethyl ethel	HCl	0.14(2)	0.09(4)	21(4)%
	HBr	0.68(2)	0.12(4)	77(4)%
	H_2SO_4	0.92(2)	0.14(4)	103(4)%
methanol	HCl	0.43(2)	0.43(4)	72(4)%
	HBr	0.50(2)	0.32(4)	74(4)%
	H_2SO_4	0.52(2)	0.41(4)	83(4)%

References

1. E. E. vanTamelen, G. Boche, S. W. Ela and R. B. Fechter, J. Am. Chem. Soc., 89, 5707 (1967).
2. E. E. vanTamelen and B. Abermark, J. Am. Chem. Soc., 90, 4492 (1968).
3. E. E. vanTamelen, R. B. Fechter, S. W. Schueller, G. Boche, R. H. Greely and B. Abermark, J. Am. Chem. Soc., 91, 1151 (1969).
4. E. E. vanTamelen and D. A. Seeley, J. Am. Chem. Soc., 91, 5194 (1969).
5. E. E. vanTamelen, R. B. Fechter and S. W. Schneller, J. Am. Chem. Soc., 91, 7196 (1969).
6. E. E. vanTamelen, D. Seeley, S. Schneller, H. Rudler and W. Cretney, J. Am. Chem. Soc., 92, 5251 (1970).
7. E. E. vanTamelen, Accounts Chem. Res., 3, 361 (1970).
8. G. Henrici-Olivé and S. Olivé, Angew. Chem. Internat. Edit., 8, 650 (1969).
9. G. Henrici-Olivé and S. Olivé, Angew. Chem, 79, 898 (1967).
10. G. Henrici-Olivé and S. Olivé, Angew. Chem. Internat. Edit., 6,873 (1967).
11. G, Henrici-Olivé and S. Olivé, Angew. Chem., 80, 398 (1968).
12. G. Henrici-Olivé and S. Olivé, Angew. Chem. Internat. Edit., 7, 386 (1968).
13. Yu. G. Borodko, M. O. Broitman, L. M. Kachapina, A. E. Shilov and L. Yu. Ukhin, Chem. Commun., 1185 (1971).
14. M. O. Broitman, N. T. Denisov, N. I. Shuvalova and A. E. Shilov, Kinet. Katal., 12, 504 (1971).
15. E. E. vanTamelen and H. Rudler, J. Am. Chem. Soc., 92, 5253 (1970).
16. N. T. Denisov, V. F. Shuvalov, N. I. Shuvalova, A. K. Shilova and A. E. Shilov, Dolk. Akad. Nauk. SSSR, 195, 879 (1970).
17. N. T. Denisov, E. I. Rudshtein, N. I. Shuvalova, A. K. Shilova and A. E. Shilov, Dolk. Akad. Nauk. SSSR, 202, 612 (1971).
18. N. T. Denisov, N. I. Shuvalova and A. E. Shilov, Kinet. Katal., 14, 1325 (1973).
19. N. T. Denisov, G. G. Terekhina, N. I. Shuvalova and A. E. Shilov, Kinet. Katal., 14, 939 (1973).
20. L. A. Nikonova, S. A. Isaeva, N. I. Pershikova and A. E. Shilov, J. Molecular Catal., 1, 367 (1975/1976).
21. N. P. Luneava, L. A. Nikonova and A. E. Shilov, Reaction Kinet. and Catal. Let., 5, 149 (1976).
22. J. Chatt, G. A. Heath and G. J. Leigh, JCS Chem. Comm., 444 (1972).
23. J. Chatt, G. A. Heath and R. L. Richards, JCS Chem. Comm., 1010 (1972).
24. G. A. Heath, R. Mason and K. M. Thomas, J. Am. Chem. Soc., 96, 259 (1974).
25. J. Chatt, J. Organometal. Chem., 100,17 (1975).

26. J. Chatt, A. J. Pearman and R. L. Richards, J. Organometal. Chem., _101_, 645 (1975).

27. J. Chatt, G. A. Heath and R. L. Richards, JCS Dalton, 2074 (1974).

28. J. Chatt, A. J. Pearman and R. L. Richards, Nature, _253_, 39 (1975).

29. C. R. Brûlet and E. E. vanTamelen, J. Am. Chem. Soc., _97_, 911 (1975).

30. A. A. Diamantis, J. Chatt, G. J. Leigh and G. A. Heath, J. Organometal. Chem., _84_, C11 (1975).

31. V. W. Day, T. A. George and S. D. Allen Iske, J. Am. Chem. Soc., _97_, 4127 (1975).

32. M. E. Vol'pin and V. B. Shur, Nature (London), 209, 1236 (1966); Dokl. Akad. Nauk. SSSR, 1102 (1964).

33. Y. G. Borod'ko, I. N. Ivleva, L. M. Kachapina, S. I. Salienko, A. K. Shilova, and A. E. Shilov, J. Chem. Soc. Chem. Commun., 1178 (1972).

34. J. E. Bercaw, R. H. Marvich, L. G. Bell and H. H. Brintzinger, J. Am. Chem. Soc., _94_, 1219 (1972).

35. C. Ungurenasu and E. Streba, J. Inorgan. Nucl. Chem., _34_,3753 (1972).

36. J. H. Teuben, J. Organometal. Chem., _57_, 159 (1973).

37. F. W. Van Der Weij and J. H. Teuben, J. Organometal. Chem., _105_, 203 (1976).

38. A. E. Shilov, A. K. Shilova and E. F. Kvashina, Kinet. Katal., _10_, 1402 (1969).

39. A. E, Shilov, E. F. Kvashina and T. A. Vorontsova, Chem. Commun., 1590 (1971).

40. G. P. Pez, J. Am. Chem. Soc., _98_, 8072 (1976).

41. G. P. Pez and S. C. Kwan, J. Am. Chem. Soc., _98_, 8079 (1976).

42. J. E. Bercaw and H. H. Brintzinger, J. Am. Chem. Soc., _93_, 2046 (1971).

43. J. E. Bercaw, J. Am. Chem. Soc., _96_, 5087 (1974).

44. J. M. Manriquez and J. E. Bercaw, J. Am. Chem. Soc., _96_, 6229 (1974).

45. T. C. McKenzie, R. D. Sanner, and J. E. Bercaw, J. Organometal. Chem., _102_, 457 (1975).

46. R. D. Sanner, D. M. Duggan, T. C. McKenzie, R. E. Marsh and J. E. Bercaw, J. Am. Chem. Soc., _98_, 8358 (1976).

47. D. Sellman, Angew. chem. Internat. Edit., _14_, 639 (1974).

48. J. E. Bercaw, E. Rosenburg and J. D. Roberts, J. Am. Chem. Soc., _96_, 612 (1974).

49. R. D. Sanner, J. M. Manriquez, R. E. Marsh and J. E. Bercaw, J. Am. Chem. Soc., _98_, 8351 (1976).

50. J. M. Manriquez, D. R. McAlister, E. Rosenberg, A. M. Shuller, K. L. Williamson, S. I. Chan and J. E. Bercaw, J. Am. Chem. Soc., to be submitted.

51. J. M. Manriquez, R. D. Sanner, R. E. Marsh and J. E. Bercaw, J. Am. Chem. Soc., _98_, 3042 (1976).

52. J. M. Manriquez, R. A. Bell and J. E. Bercaw, unpublished.

ASYMMETRIC HOMOGENEOUS CATALYSIS BY CHIRAL TRANSITION METAL COMPLEXES

Piero Pino and Giambattista Consiglio

Swiss Federal Institute of Technology, Department of Industrial and Engineering Chemistry, Universitätstrasse 6, 8092 Zurich, Switzerland

Asymmetric catalysis is the most common way to synthesize single antipodes in biological processes.[1] It would also be the most convenient way to prepare optically active compounds in the laboratory since in principle very small amounts of antipodes could be used to synthesize an unlimited quantity of a chiral compound having a specific chirality. In spite of this, asymmetric catalysis, with the exception of enzymatic catalysis,[1] was very little investigated until the early sixties.

It was observed long ago[2] that optically active cyanohydrins can be obtained from aldehydes and hydrogen cyanide in the presence of optically active bases, but this was not further investigated until 1954.[3] Other asymmetric catalytic reactions[4] have been studied only rather recently such as, for example, the addition of alcohols to phenylmethylketene[5] or the addition of bromine to olefins[6] (both catalyzed by alkaloids).

In 1954 prochiral olefins* were polymerized[7] stereospecifically to yield isotactic polymers. This synthesis does not yield optically active compounds, but can be considered as an asymmetric synthesis[10] catalyzed by racemic transition metal complexes.[11] Using heterogeneous catalysts a very high excess of a single diasteromer is obtained.[11]

---

* Actually vinylpolymers containing some isotactic diastereomers were obtained between 1948[8] and 1950,[9] but they were not isolated and their structures were not correctly established.

In 1956 heterogeneous hydrogenation catalysts, the surface of which were modified by optically active compounds, were used for the first time.[12] Rather low[13] optical yields* were obtained with many substrates; however, very high optical yields have been recently achieved[17] with this type of catalysts.

Seven years later the first report on homogeneous asymmetric catalysis by a transition metal complex appeared[18] and it described the polymerization of trans-1,3-pentadiene with a chiral Ziegler-Natta catalyst; a similar catalytic system was used for the asymmetric cyclooligomerization[19] of the same substrate.

At this time asymmetric catalysis is a very large and rapidly developing field. In this review we shall deal only with homogeneous catalysis by chiral transition metal complexes. We shall first review catalytic reactions which have been investigated up to now, then the type of catalysts which can be used, and the chiral ligands, which have produced the highest degrees of asymmetric induction.**

---

* The optical yield is defined as the ratio of the optical purity of the product to the optical purity of the optically active component in the reaction.[14] Optical purity has been defined as the ratio of the specific rotation of a substance to the specific rotation of the pure enantiomer.[15] The value of optical purity is equal to the value of the enantiomeric excess[16] or enantiomeric purity[15] (defined as the excess of one enantiomer over the other) only if a linear relationship between rotation and composition exists. The enantiomeric excess defines exactly the enantiomeric composition of a chiral substance; however, it is seldom determined. Normally, from a reliable value of the rotation of a pure enantiomer the optical purity can be determined. However, it is not correct to speak about enantiomeric excess on the basis of optical rotation without checking the aforementioned linear relationship.

** Asymmetric induction defines the influence of a chiral entity on a prochiral[20] moiety to form a new chiral entity, stereoisomers being formed in different amounts.[21] For a complete definition of the phenomenon, type and extent of asymmetric induction must be given. The type specifies the prevailing configuration of the new chiral entity and can be conveniently indicated following the Cahn, Ingold and Prelog nomenclature.[22] The extent of asymmetric induction can only be indicated if the optical purity of the inducing chiral entity is known. In this case the extent of the asymmetric induction corresponds to the optical yield; if a linear relationship between optical rotation and enantiomeric excess can be assumed (for diastereoface differentiating reactions[23] it corresponds to the diastereomeric excess). If the optical purity (and if necessary diastereomeric purity) of the inducing chiral entity is unknown, only a minimum asymmetric induction can be indicated, since the asymmetric induction is larger or equal to the optical purity of the product.

Finally we shall briefly consider the main open questions in asymmetric catalysis and the present trends of the research in the field.

This paper is not an exhaustive account of all the work done in the field of asymmetric catalysis by homogeneous transition metal complexes; we hope, however, that it can give to the newcomers in the field an idea of the main research lines which have been followed up to now.

Type of Asymmetric Reactions Catalyzed by Chiral Transition Metal Complexes

Since in most cases the mechanism of the catalytic reaction is unknown, the reactions will be classified by taking into account only the substrate and products.[*]

The most common asymmetric reaction successfully investigated is the enantioface discriminating synthesis[23] (Scheme 1). This type of reaction, once improperly called an enantioselective reaction,[13] includes, among others, asymmetric hydrogenation,[28] hydrocarbonylation,[29] hydrosilylation,[30] ethylene-olefins-codimerization[31] (Table 1). Most of the carbene addition[**] to double bonds[59-74] are further examples of enantioface discriminating reaction.

---

[*] In some cases and particularly when the mechanism of the catalytic reaction is not understood, the classification of a catalytic asymmetric transformation according to the type of the substrates and products can be misleading. A typical case is the asymmetric cross-coupling reaction involving an aliphatic racemic Grignard reagent.[24, 25, 26] In this case the reaction is formally a kinetic resolution, followed by rapid racemization of the non-reacted antipode.[27] However, taking into account that the catalytic process might in some cases involve the formation of a π-complex[24] the reaction might also be considered to be an enantioface discriminating reaction.

[**] When cyclopropane derivatives are synthesized from diazocompounds and olefins, optically active products can also be obtained from non-prochiral olefins. A typical example is the synthesis of optically active ethyl 2,2-diphenylcyclopropane carboxylate from 1,1-diphenylethylene and ethyl diazoacetate.[73] In this case the prochiral reaction educt is the carbon atom bound to the diazogroup and bearing two different substituents. After elimination of the nitrogen molecule, the carbenoid atom can bond to the two nonequivalent carbon atoms of the olefinic substrate in two ways yielding two enantiomers, one way being preferred. In this respect the reaction can be considered to be an enantiosites[23] discriminating reaction.

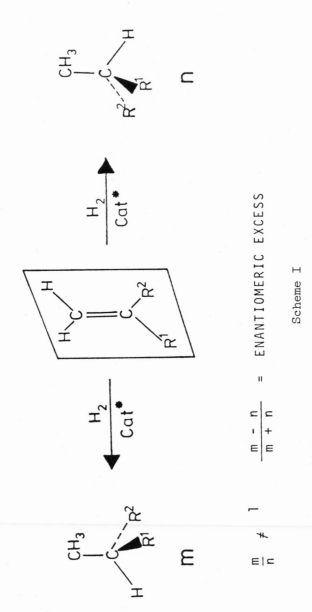

ENANTIOFACE DISCRIMINATING SYNTHESIS

$\dfrac{m - n}{m + n}$ = ENANTIOMERIC EXCESS

$\dfrac{m}{n} \neq 1$

Scheme I

Table 1. Some enantiofaces differentiating reactions reported in the literature[a]

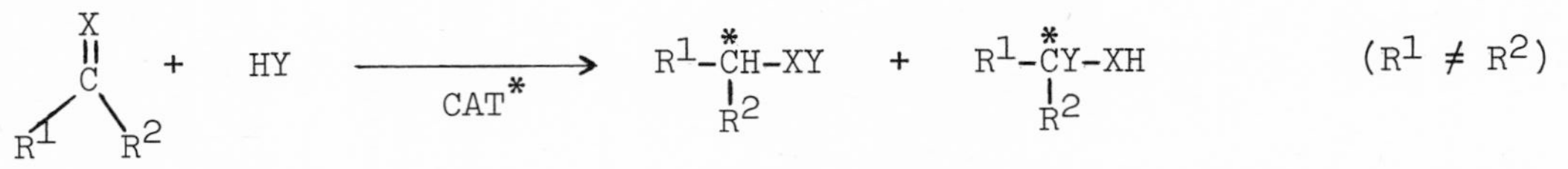

$$X{=}C\begin{smallmatrix}R^1\\R^2\end{smallmatrix} + HY \xrightarrow{CAT^*} R^1{-}\overset{*}{C}H{-}XY \atop R^2 + R^1{-}\overset{*}{C}Y{-}XH \atop R^2 \quad (R^1 \neq R^2)$$

Y	$=C\begin{smallmatrix}R^3\\R^4\end{smallmatrix}$	=O	=NR	=NOH
-H	32-44	45-53	50, 54	49
-COR	55-60			
-SiR'R''R'''	61	32, 62-66	54, 67	
-CH=CH$_2$	68			
M[b]				

a. Figures correspond to references in which the corresponding reactions are described.
b. M indicates a transition metal atom bound to other substituents.
c. R3 and R4 can be H or other substituents. Depending on the nature of R3 and R4 two asymmetric carbon atoms can form during the reaction.

It must be emphasized that in the cases where a reagent of the type HX (X≠H) is added to a C_s olefin, two isomers are formed, the amount of which is determined by regioselectivity. In this case if we are concerned with an enantioface discriminating reaction, the type and extent of asymmetric induction will depend not only on the enantioface discrimination, but also, and sometimes predominantly, on regioselectivity.[75] Asymmetric hydroformylation,[55, 57, 76] and hydrocarboxylation[56] have been obtained also with C_{2v} olefins. In this case the preferential attack of the COX group (Scheme 2a) to either prochiral face of the two equivalent trigonal carbon atoms can rise from the side of the olefin initially attacked and/or, if only the attack to one of the two identical faces of the olefin is considered, from regioselectivity. As a matter of fact, regioselectivity yields in this case two enantiomers and not two structural isomers.

Finally a very peculiar case is represented by the asymmetric hydroformylation,[75] hydrocarboxylation[75] or hydrovinylation[68] of norbornene. Both with chiral or achiral catalysts diastereoface selection* for this substrate is almost complete only exo products (>97%) being formed. In this case the two possible regioisomers resulting from the attack to the exo face of the substrate are enantiomers, and therefore asymmetric induction is connected to only regioselectivity (Scheme 2b).

Regioselectivity seems to be an important factor also in the asymmetric (2+2) cross addition of olefins.[77]

A second important type of asymmetric reaction which has been successfully carried out using chiral transition metal complexes is the kinetic resolution of racemic compounds.[78] This name for an asymmetric transformation puts the emphasis on the non-transformed substrate. However, as new chiral compounds are sometimes synthesized during the kinetic resolution and catalyst discriminates be-

---

* Diastereoface selection[23] implies the formation of a chiral moiety in a substrate which already contains at least one chirality center. Except in peculiar cases as norbornene, which possesses a meso structure, diastereoface discriminating reactions have mostly been carried out using achiral catalysts and will not be treated in this paper.

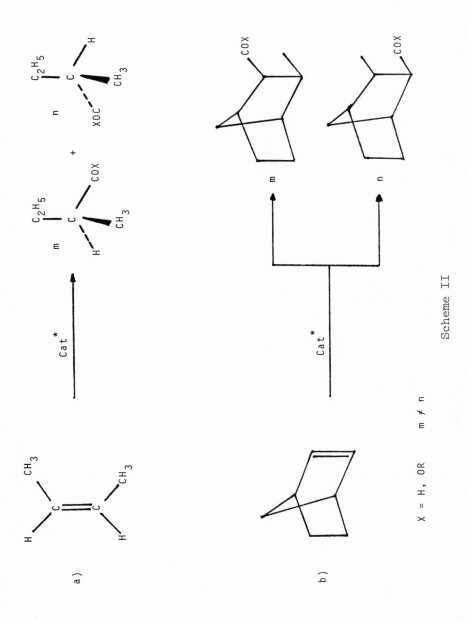

Scheme II

tween the two antipodes, both the terms "antipodes discriminating reaction"[23] or "enantioelective reaction"[68] seems more appropriate.*
Furthermore, the antipodes discriminating reactions are not necessarily kinetic resolutions as shown by the recently described[79] regioselectivity controlled asymmetric hydroformylation of racemic olefins.

Table 2 presents some antipodes discriminating reactions by transition metal complexes.

A further type of catalytic asymmetric reaction is an enantiotopic group discriminating substitution.[23] Only one example of this reaction is known up to now, the substitution of a hydrogen atom in a prochiral silane $(R_1R_2SiH_2; R_1 \neq R_2)$, which is catalyzed by chiral rhodium complexes.[95]

Another interesting type of catalytic asymmetric transformation is the conformationally controlled nickel catalyzed coupling reaction.[96] Optically active 2,2'-dimethyl-1,1'-binaphthyl has

---

* In the case of the synthesis of compounds containing more than one asymmetric carbon atom from racemic substrates containing only one asymmetric carbon atom, there are two types of antipode discriminating reactions which lead to two different types of diastereomers:

$$(S) + (R) + (S)' + (R)' \longrightarrow \begin{array}{l} (SS') + (RR') \\ (SR') + (RS') \end{array}$$

$$(S) + (R) + (S)' + (R)' \underset{cat^*}{\overset{cat^*}{\longrightarrow}} \begin{array}{l} SS' + (R) + (R)' \\ SR' + (R) + (S)' \end{array}$$

We think that, as proposed for the polymerization of α-olefins[11] the term "stereoselective reaction" should be reserved for the first type of the above reactions which leads to racemic compounds. The first type can be catalyzed by racemic catalysts, while the second one needs the presence of an optically active catalyst.

Table 2.   Some antipodes differentiating reactions reported in the literature[a]

Type of Reaction	Substrate					
	$R^*\text{-}C{=}C$ [b]	R*–OH[b]	$CH_2\text{-}\overset{*}{CH}\text{-}CH_3$ (O epoxide)	$CH_3\text{-}\overset{*}{CH}\text{-}CH_2Cl$ (OH)	$R^*\text{-}C{=}O\text{-}R$ [b]	$CH_3\text{-}\overset{*}{CH}$ (CO–O / NH–CO)
Hydrogenation	80				81	
Dehydrogenation		82-86				
Hydrocarbonylation[c]	57, 79, 80					
Hydrogen shift	87-89		90, 91			
Hydrogen transfer	92					
Dehydrohalogenation				93		
Hydrovinylation[d]	68					
Polymerization						94

a Figures correspond to references in which the reactions are described.
b R* means a substituent containing at least one chirality center.
c Hydrocarbonylation includes all the reactions in which an hydrogen atom and a CO molecule are added to a double bond.
d We think that analogously to c) the term hydrovinylation conveniently describes a reaction in which a hydrogen atom and a vinyl group are added to a double bond. These reactions are in general indicated as codimerizations of an olefin with ethylene.

been prepared from 1-bromo-2-methylnaphthalene and 2-methyl-1-naph-thylmagnesiumbromide.

From the above discussion it appears that asymmetric homogene-ous catalysis can have broad applications even if in most cases the extent of the asymmetric induction is indeed small and the prospects for a substantial increase in the enantiomeric excess obtainable is generally not very good. This difficulty is mainly due to a lack of knowledge about the mechanism by which asymmetric induction takes place, as we shall discuss later.

Asymmetric Catalysts

Since the catalytic action of coordination complexes is con-nected with the formation and the splitting of metal-substrate bonds,[97] the most efficient asymmetric synthesis should be achieved when complexes containing "asymmetric" or "chiral" metal atoms are used.[98-100] As a matter of fact, this would minimize distance be-tween the inducing asymmetric center and the asymmetric moiety be-ing formed. However, coordination complexes containing metal-to carbon bonds (organometallic complexes) are in general configura-tionally labile.[98-100] In order to improve the configurational stability tightly bonded ligands (such as the cyclopentadienyl group) must be used.[101] However, this generally lowers the cata-lytic activity of the complex by a great extent.

Chiral complexes for asymmetric catalysis can also be obtained using chiral ligands.[99, 100] If a chiral metal atom is bound to a ligand containing a single non-racemic chirality center (e.g. an asymmetric carbon atom), two diastereoisomers must exist (see, for example, the complexes of Scheme 3). The two diastereoisomers will have different free energies of formation, and even if the rate at which the asymmetric metal atom inverts its configuration is high, either diastereomer (only one antipode of each, if the ligand is optically pure and the chiral center present in the ligand is con-figurationally stable under reaction conditions) will prevail. Hence, asymmetric metal atoms of one chirality will prevail.

Until recently nothing was known about epimerization equilibria for organometallic complexes containing one asymmetric metal atom. However, in a paper concerning[102] complexes containing an asymmetric molybdenum or tungsten atom and one asymmetric carbon atom (Scheme 3) it has been shown that the two diastereomers can have sufficient-ly different thermodynamic stabilities so that one of the diastereo-mers will largely prevail (87:13). The fact that some ligands can cause a very high extent of asymmetric induction in hydrogenation can indicate that in some cases one of the diastereomers largely prevails. In fact, if the chirality of the metal atom plays an

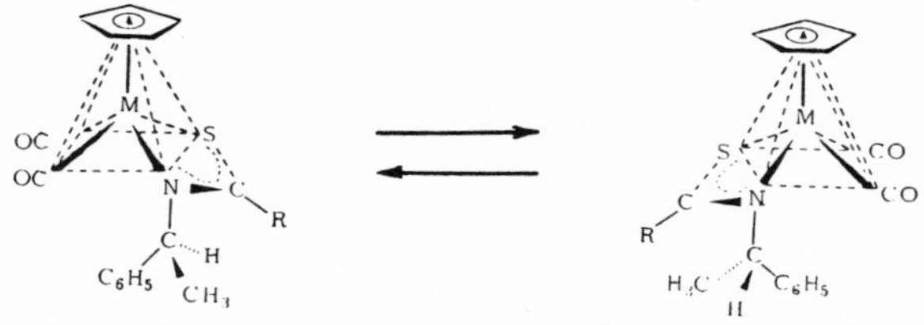

R \\ M	-H	-CH$_3$	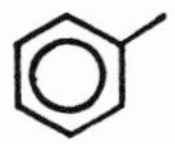 (C$_6$H$_5$)	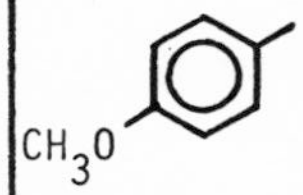 (CH$_3$O-C$_6$H$_4$)	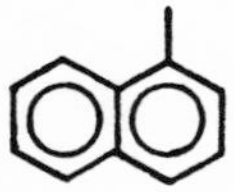 (naphthyl)	 (methyl-naphthyl)
Mo	50 : 50 (40°C)	69 : 31 (50°C)	77 : 23 (60°C)	75 : 25 (55°C)	87 : 13 (25°C)	78 : 22 (55°C)
W	50 : 50 (40°C)	75 : 25 (40°C)	83 : 17 (50°C)			

Diastereomeric ratio [(+)/(−)] at equilibrium conditions in CDCl$_3$ as the solvent[102]. (+) and (−) indicate the diastereomer having positive, respectively negative rotation at 365 nm.

Scheme III

important role in asymmetric induction, it seems reasonable to as-
sume that high extents of asymmetric inductions must be connected
with a large prevalence of one antipode of a single diastereomer of
the organometallic catalytic complex.

Ligands of different nature such as a phosphine containing an
asymmetric phosphorus atom,[103] a diphosphine containing two asymmetric carbon atoms[104] and a ferrocenyldiphosphine[36] all cause a
very high asymmetric induction in the hydrogenation of the same
substrate. This fact could be explained by assuming that for different ligands, configurationally stable under the conditions used
for the catalytic reaction the epimerization equilibrium in the complex is similarly displaced, giving rise to similar prevalences of
the same diastereomer.

Although in our opinion the presence of a chiral metal atom is
an important factor in asymmetric catalysis by transition metal complexes, asymmetric induction can occur also in square planar complexes not containing chiral metal complexes. As a matter of fact,
stereoelective complexation of prochiral olefins[105, 106] and of
chiral α-olefins[107] in cis- and trans-dichloro (1-phenylethylamine)-
(olefin)platinum (II) complexes* has been known for several years.
Finally the hypothesis of a direct interaction between asymmetric
ligand and substrate (vide infra) cannot be entirely discarded.

Asymmetric Ligands

The investigation of asymmetric catalysis by transition metal
complexes has been carried out up to now using metal complexes containing chiral ligands. In Table 3 the number of papers in which
the use of the most common optically active ligands are reported.
Although the survey of literature data (limited to the end of 1976)
is probably not complete and the patent literature has not been
considered in detail, the table should give an idea of the present
situation in the field. The ligand most used in asymmetric catalysis is by far the DIOP introduced for the first time by Kagan;[104]
less frequently used are phosphines containing chiral phosphor
atoms.[28, 34] Phosphines containing chiral hydrocarbon groups are
easily prepared and have been used in six different asymmetric re-

---

* In this case it is probable that because of the position of the
conformational equilibrium in the asymmetric ligand, and/or because of secondary interaction between some groups of the ligands
(e.g., a phenyl group) and the metal, the space above and below
the coordination plane is unequally occupied simulating an higher
coordination around the metal, which then behaves as a chirality
center.

Table 3. Optically active ligands most used for asymmetric catalytic reactions.[a,b]

Reaction \ Asymmetric Ligand	DIOP and analogous	*P‹	P-R*	P-OR*	Ferrocenyl phosphines	Alcaloids	Shiff bases	Alkoxy groups
Reduction	20	13	5	1	2	9	1	1
Hyrdocarbonylation	12	3	2		1			
Hydrosilylation	11	13	2		1			
Cross-coupling	3		3					
Hydrovinylation								
Substitution (Si)	4							
Isomerization		1						1
Hydrogen shift							3	
Polymerization			2					3
Oligomerization			3					1
Dehydrogenation								
Carbenoid reaction							6	
Hydrogen transfer							2	
Dehydrohalogenation				1			1	
Hydroamination[c]								

a Figures correspond to the number of published papers in which the use of each asymmetric ligand has been described.
b Data are based on the most significant scientific papers and patents appeared up to the end of 1976.
c Addition of an hydrogen and a $-N\langle^R_R$ group to a double bond.

actions.[42, 58, 68, 85, 108, 109] Nitrogen compounds, particularly alkaloids[53] and Schiff bases[29] have been also used with less encouraging results. Among the other types of ligands, phosphites[71] containing chiral groups and chiral alkoxy groups[18] have also some significance. There is increasing interest for the last ones in particular.[94]

Other asymmetric ligands which are more seldom used are carboxylic acids,[89] amines,[110] amides,[35] ethers,[111] dioximes[73] and sulfoxides.[41]

The highest extent of asymmetric induction has been obtained in hydrogenation when DIOP[33] other diphosphines,[44, 112] phosphines containing chiral phosphor atoms[103] or ferrocenyl phosphines[36] have been used. However, degrees of asymmetric inductions over 50% have been achieved also in hydrosilylation,[54, 63-67, 113] carbenoid reactions,[73, 74] cross-coupling reaction,[26] hydrocarbonylation,[56] hydrovinylation[68] and reduction of keto groups.[52] Up to now the choice of ligands has been mainly empirical and attempts to find relationships between ligand structure and extent of asymmetric induction obtainable in catalytic reactions have failed.[42]

This situation does not favour the practical application of asymmetric catalysis in organic synthesis, since optimization of the extent of asymmetric induction must be done empirically and requires a very large experimental effort. The investigation of diastereomeric equilibria in chiral complexes containing chiral ligands and chiral metal atoms might largely contribute to a better understanding of at least one of the numerous factors influencing asymmetric induction.

Main Open Questions in Asymmetric Catalysis by Transition
Metal Complexes

The most important questions in asymmetric catalysis both from a theoretical and from a synthetic point of view are the determination of the step or steps in which asymmetric induction takes place and the clarification of the mechanism by which asymmetric induction takes place.

Catalytic asymmetric syntheses are in general multi-step processes[97] and the asymmetric induction is determined before the first substantially irreversible step. An appreciable reversibility without reaching equilibrium conditions of all the steps preceding the first irreversible step is the necessary condition that these steps contribute to determine the asymmetric induction.[114]

The present knowledge about the step or steps regulating asymmetric induction is shown in Table 4. Only in the case of hydrocarbonylation has it been possible to accumulate enough experimental evidence to show the step before which, at which or after which, the

Table 4. Step of the catalytic process which can partake to the control of the asymmetric induction.

A Substrate-catalyst complex formation
B Diastereomeric catalyst-substrate complexes equilibrium
C Insertion reaction
D Equilibrium between diastereomeric metal alkyls
E Other intramolecular substrate-catalyst complexes isomerization
F Reductive elimination

Reaction	Catalyst	Ligand	Substrate	Step	Direct or indirect experimental evidence	No experimental evidence
Hydrogenation	Rh	(-)-DIOP	$\phi-\overset{}{C}=C\overset{COOR^{2a,b}}{\underset{R^3}{}}$, R^1	B, C	+ 116	
Hydrogenation	Rh	(-)-1-phenylethyl-formamide		B, C, D	+ 35	
Hydroformylation	Rh	(-)-DIOP	trans-butene, 1-butene, cis-butene	A, B, C	+ 55	
Hydroformylation	Pt	(-)-DIOP		D, E, F	+ 57	
Hydrocarboxylation	Pd	(-)-DIOP	2-methyl-1-butene, 2-methyl-2-butene	A	+ 115	
Hydrovinylation	Ni	(-)-Pr^{i}P(Menthyl)$_2$	norbornene	C		+ 68
Hydrosilylation	Rh	BzMePhP*	$R^1\overset{}{\underset{R^1}{}}C=O$	C		+ 66
Cross-coupling	Ni	(-)-DIOP	$\phi-CH-MgBr$, $CH_3+CHCl=CH_2$	A, F		+ 24

a Rh-(-)-DIOP, cis and trans isomer: $R^1 = R^2 = H$; $R^3 = NHCOC_6H_5$
b Rh-(-)-1-PEF, cis and trans isomer: $R^1 = R^2 = CH_3$; $R^3 = H$

asymmetric induction takes place. In particular in the hydrocarboxylation of 2-phenyl-1-butene or 2-methyl-1-butene[115] has it been
shown that asymmetric induction occurs, at least in part, in the
first step.

The experiments have shown that no generalization is possible.
In the rhodium catalyzed hydroformylation and in palladium catalyzed hydrocarboxylation of butenes using DIOP as asymmetric ligand
the asymmetric induction must take place, at least with the substrates used, in large part before or during the metal-alkyl complex formation[55, 56] which is practically irreversible under the
conditions used. In the platinum catalyzed hydroformylation of the
same substrates using the same ligand the asymmetric induction either
reflects the equilibrium between the two diastereomeric platinum
alkyl complexes or takes place after this step.[57]
Although there is not much experimental evidence, also the
structure of the substrate and of the ligand may in principle influence the step or steps in which asymmetric induction takes place.
However, using two different ligands used in asymmetric hydroformylation catalyzed by rhodium, the step controlling asymmetric induction
reamins the same.[76]

In asymmetric hydrogenation using rhodium catalysts in the
presence of DIOP as the ligand, the different optical yield obtained
with (Z) and (E) isomers has been interpreted to indicate that asymmetric induction takes place during or before the rhodium alkyl formation.[116] All the other indications are based only on speculation
connected with the reaction mechanism.

It is difficult to propose a sound explanation for the manner
in which one or more reaction steps can control the type and extent
of asymmetric induction. Attempts to interpret the results on the
basis of models of reaction intermediates have been made for hydroformylation[55] and for hydrogenation[117] of olefins, for hydrosilylation of ketones,[66, 117] and for hydrocarboxylation.[118]
The main factors influencing asymmetric induction which have
been considered up to now are the following:
a.) The optical purity of the chiral metal atoms,[55] if the
catalyst contains chiral metal atoms.[119]
b.) The stereochemistry of the catalytic complexes.[37, 57, 119]
c.) The differences between the repulsive forces existing between the substrate and the catalytic complex,[55] when either prochiral face is coordinated.
d.) The presence of attractive or repulsive forces between the
substrate and the chiral ligand.[35, 36, 120]
At the present, however, it is difficult to judge the relative importance of the above factors.

Present Trends in Research on Asymmetric Catalysis

The most attractive direction for studies on asymmetric catalysis by transition metal complexes is certainly the search for new
asymmetric reactions and for new and more efficient catalytic systems and in particular for new and more efficient chiral ligands
which would improve the obtainable enantiomeric excess. However,
there are further directions which promise to be very important to
the future development of asymmetric catalysis.

The first one, as we have already mentioned, is the investigation of diastereomeric equilibria in transition metal complexes
containing asymmetric ligands. This research requires first the
identification of suitable model systems where the epimerization of
the complexes and of degree of optical purity of the asymmetric
metal atom could be monitored. In this connection the investigation
of the factors controlling stereoselectivity in reactions involving
organometallic complexes[83] can contribute to the determination of the
relative free energies of formation of organometallic diastereomeric
products, which are important intermediates in asymmetric catalysis.

Another interesting direction is the investigation of the influence of the reaction variables on the asymmetric induction. The
importance of such investigation is twofold: improvement of the
enantiomeric excess can be achieved and mechanistic information can
be obtained. The influence of some reaction variables has been investigated in some details only in the case of the rhodium catalyzed
hydrocarboxylation. For other asymmetric reactions practically nothing is known.

Finally the systematic investigation of attractive forces between substrate and catalyst should be carried out. In fact attractive forces rather than repulsive forces should permit to reach
easily $\Delta\Delta G^{\ddagger}$ of 2000-3000 calories in the step controlling asymmetric
synthesis, which would allow to achieve an enantiomeric excess of
practical importance for the synthesis of single antipodes.

Financial support by the Schweizerischer Nationalfonds zur
Förderung der wissenschaftlichen Forschung is kindly acknowledged.

References

1. J. B. Jones, J. F. Beck, "Application of Biochemical Systems in
Organic Synthesis" Part I, J. B. Jones, C. J. Sih, D. Perlman, Ed.,
Wiley, New York, N.Y., 1976, p. 107-401.
2. G. Bredig, P. S. Fiske, Biochem. Z., _46_, 7 (1912).
3. V. Prelog, M. Wilhelm, Helv. Chim. Acta, _37_, 1634 (1954).
4. For a review see: H. Pracejus, Fortsch. Chem. Forsch., _8_,
493 (1967).

5. H. Pracejus, Ann., <u>634</u>, 9 (1960).
6. G. Berti, A. Marsili, Angew. Chem., <u>75</u>, 1026 (1963).
7. G. Natta, P. Pino, P. Corradini, F. Danusso, E. Mantica,
G. Mazzanti, G. Moraglio, J. Amer. Chem. Soc., <u>77</u>, 1708 (1955).
8. C. E. Schildknecht, S. F. Gross, H. R. Davidson, J. M. Lambert,
A. Zoss, Ind. Eng. Chem., <u>40</u>, 2108 (1948).
9. A. A. Morton, Ind. Eng. Chem., <u>42</u>, 1488 (1950).
10. J. D. Morrison, H. S. Mosher, "Asymmetric Organic Reactions,"
Prentice-Hall, Inc., Englewood Cliffs, N.J., 1971, p. 425.
11. For a review see: P. Pino, U. W. Suter, Polymer, <u>17</u>, 977 (1976).
12. S. Akabori, Y. Izumi, Y. Fuji, S. Sakurai, Nature, <u>178</u>, 323
(1956).
13. For a review see: Y. Izumi, Angew. Chem., <u>83</u>, 956 (1971).
14. E. L. Eliel, "Stereochemistry of Carbon Compounds," International
Student Edition, McGraw Hill, New York, 1962, p. 76.
15. M. Raban, K. Mislow, Topics in Stereochemistry, N. L. Allinger
and E. L. Eliel Eds., Vol. 2, 1967, p. 199.
16. See ref. 10, p. 10ff.
17. Y. Orito, S. Niwa, S. Imai, Yuki Gosei Kagaku Kyokai Shi, <u>34</u>,
672 (1976); C. A., <u>86</u>, 43148 (1977).
18. G. Natta, L. Porri, S. Valenti, Makromol. Chem., <u>67</u>, 225 (1963).
19. J. Furukawa, T. Kakuzen, H. Morikawa, R. Yamamoto, O. Okuno,
Bull. Chem. Soc. Japan, <u>41</u>, 155 (1968).
20. K. R. Hanson, J. Amer. Chem. Soc., <u>88</u>, 2731 (1966).
21. See ref. 14, p. 68.
22. R. S. Cahn, C. Ingold, V. Prelog, Angew. Chem., <u>78</u>, 413 (1966).
23. This nomenclature has been proposed for the first time by some
authors in a Japanese book: "Chemistry of Asymmetric Reactions"
Kagaku Sosetsu No. 4, 1974 (see, for example, C. A., <u>82</u>, 30459
(1975)).
24. G. Consiglio, C. Botteghi, Helv. Chim. Acta, <u>56</u>, 460 (1973).
25. Y. Kiso, K. Tamao, N. Miyake, K. Yamamoto, M. Kumada, Tetrahedron Lett., 3 (1974).
26. T. Hayashi, M. Tajika, K. Tamao, M. Kumada, J. Amer. Chem. Soc.,
<u>98</u>, 3718 (1976).
27. F. R. Jensen, K. L. Nakamaye, J. Amer. Chem. Soc., <u>88</u>, 3437
(1966) and references therein.
28. W. S. Knowles, M. J. Sabacky, Chem. Comm., 1445 (1968).
29. C. Botteghi, G. Consiglio, P. Pino, Chimia, <u>26</u>, 141 (1972).
30. K. Yamamoto, T. Hayashi, M. Kumada, J. Amer. Chem. Soc., <u>93</u>,
5301 (1971).
31. B. Bogdanović, B. Henc, H. G. Karman, H. G. Nuessel, D. Walter,
G. Wilke, Ind. Eng. Chem., <u>62</u> (12), 35 (1970).
32. W. S. Knowles, M. J. Sabacky, B. D. Vineyard, D. J. Weinkauff,
J. Amer. Chem. Soc., <u>97</u>, 2567 (1975) and therein references.
33. H. B. Kagan, Pure Appl. Chem., <u>43</u>, 401 (1975) and therein
references.
34. L. Horner, H. Siegel, Phosphorus, <u>1</u>, 209 (1972) and therein
references.

35. P. Abley, F. J. McQuillin, J. Chem. Soc. (C), 844 (1971).
36. T. Hayashi, T. Mise, S. Mitachi, K. Yamamoto, M. Kumada, Tetrahedron Lett., 1133 (1976).
37. M. Tanaka, I. Ogata, J. C. S. Chem. Comm., 735 (1975).
38. M. Fiorini, G. M. Giongo, F. Marcati, W. Marconi, J. Mol. Cat., 1, 451 (1976).
39. S. Takeuchi, Y. Ohgo, J. Yoshimura, Chem. Lett., 265 (1973).
40. B. R. James, D. K. W. Wang, R. F. Voigt, J. C. S. Chem. Comm., 574 (1975).
41. B. R. James, R. S. McMillan, K. J. Reimer, J. Mol. Catal., 1, 439 (1976).
42. For a review, see, J. D. Morrison, W. F. Masler, M. K. Neuberg, Adv. Catal., 26, 81 (1976).
43. T. Hayashi, M. Tanaka, I. Ogata, Tetrahedron Lett., 295 (1977).
44. K. Achiwa, J. Amer. Chem. Soc., 98, 8265 (1976).
45. P. Bonvicini, A. Levi, G. Modena, G. Scorrano, J. C. S. Chem. Comm., 1188 (1972).
46. M. Tanaka, Y. Watanabe, T. Mitsudo, H. Iwane, Y. Takegami, Chem. Lett., 239 (1973).
47. J. Solodar, Chem. Techn., 421 (1975).
48. B. Heil, S. Toros, S. Vastag, L. Markó, J. Organometal. Chem., 94, C47 (1975).
49. C. Botteghi, M. Bianchi, E. Benedetti, U. Matteoli, Chimia, 29, 256 (1975).
50. A. Levi, G. Modena, G. Scorrano, J. C. S. Chem. Comm., 6 (1975).
51. R. W. Waldron, J. H. Weber, Inorg. Chim. Acta, 18, L3 (1976).
52. T. Hayashi, T. Mise, M. Kumada, Tetrahedron Lett., 4351 (1976).
53. Y. Ohgo, Y. Natori, S. Takeuchi, J. Yoshimura, Chem. Lett., 1327 (1974) and therein references.
54. H. B. Kagan, N. Langlois, T. P. Dang, J. Organometal. Chem., 90, 353 (1975).
55. P. Pino, G. Consiglio, C. Botteghi, C. Salomon, Adv. Chem. Ser., 132, 295 (1974) and references therein.
56. G. Consiglio, P. Pino, Chimia, 30, 193 (1976) and references therein.
57. G. Consiglio, P. Pino, Helv. Chim. Acta, 59, 642 (1976).
58. M. Tanaka, Y. Watanabe, T. Mitsudo, Y. Takegami, Bull. Chem. Soc. Japan, 47, 1698 (1974) and references therein.
59. R. Stern, A. Hirschauer, L. Sajus, Tetrahedron Lett., 3247 (1973).
60. A. Stefani, D. Tatone, Helv. Chim. Acta, 60, 518 (1977).
61. K. Yamamoto, T. Hayashi, Y. Uramoto, R. Ito, M. Kumada, J. Organometal. Chem., 118, 331 (1976) and references therein.
62. R. J. P. Corriu, J. J. E. Moreau, J. Organometal. Chem., 85, 19 (1975).
63. T. Hayashi, K. Yamamoto, M. Kumada, J. Organometal. Chem., 112, 253 (1976) and references therein.
64. T. Hayashi, K. Yamamoto, K. Kasuga, H. Omizu, M. Kumada, J. Organometal. Chem., 113, 127 (1976) and references therein.
65. J. Benes, J. Hetflejs, Coll. Czec. Chem. Comm., 41, 2264 (1976).

66. I. Ojima, T. Kogure, M. Kumagai, S. Horiuchi, T. Sato, J. Organometal. Chem., 122, 83 (1976) and references therein.
67. N. Langlois, T. P. Dang, H. B. Kagan, Tetrahedron Lett., 4865 (1973).
68. B. Bogdanović, B. Henc, A. Loesler, B. Meister, H. Pauling, G. Wilke, Angew. Chem., 85, 1013 (1973) and references therein.
69. H. Nozaki, H. Takaya, S. Moriuti, R. Noyori, Tetrahedron, 24, 3655 (1968).
70. R. Noyori, H. Takaya, Y. Nakanisi, H. Nozaki, Can. J. Chem., 47, 1242 (1969).
71. W. R. Moser, J. Amer. Chem. Soc., 91, 1135 (1969).
72. I. S. Lishanskii, V. I. Pomerantsev, N. G. Illarionova, A. S. Kachaturov, T. I. Vakorina, Kin. Katal., 7, 1803 (1971); C. A., 76, 3303 (1972).
73. Y. Tatsuno, A. Konishi, A. Nakamura, S. Otsuka, J. C. S. Chem. Comm., 588 (1974).
74. T. Aratani, Y. Yoneyoshi, T. Nagase, Tetrahedron Lett., 1707 (1975).
75. P. Pino et. al., in preparation.
76. M. Tanaka, Y. Ikeda, I. Ogata, Chem. Lett., 1115 (1975).
77. R. Noyori, T. Ishigami, N. Hayashi, H. Takaya, J. Amer. Chem. Soc., 95, 1674 (1973).
78. See ref. 10, p. 30ff.
79. A. Stefani, D. Tatone, P. Pino, Helv. Chim. Acta, 59, 1639 (1976).
80. G. Consiglio, unpublished results.
81. J. Solodar, D. O. S. 2.312.924; C. A., 80, 3672 (1974).
82. K. Ohkubo, K. Hirata, K. Yoshinaga, M. Okada, Chem. Lett., 183 (1976).
83. K. Ohkubo, K. Hirata, K. Yoshinaga, Chem. Lett., 577 (1976).
84. K. Ohkubo, T. Ohgushi, K. Yoshinaga, Chem. Lett., 775 (1976).
85. K. Ohkubo, T. Aoji, K. Hirata, K. Yoshinaga, Inorg. Nucl. Chem. Letters, 12, 837 (1976).
86. K. Ohkubo, K. Hirata, T. Ohgushi, K. Yoshinaga, J. Coord. Chem., 6, 185 (1977).
87. C. Carlini, D. Politi, F. Ciardelli, Chem. Comm., 1260 (1970).
88. G. Strukul, M. Bonivento, R. Ros, M. Graziani, Tetrahedron Lett., 1791 (1974).
89. G. Sbrana, G. Braca, E. Giannetti, J. C. S. Dalton, 1847 (1976).
90. H. Aoi, M. Ishimori, S. Yoshikawa, T. Tsuruta, J. Organometal. Chem., 85, 241 (1975).
91. H. Aoi, M. Ishimori, T. Tsuruta, Bull. Chem. Soc. Japan, 48, 1897 (1975).
92. L. Lardicci, G. P. Giacomelli, P. Salvadori, P. Pino, J. Amer. Chem. Soc., 93, 5794 (1971).
93. M. Ishimori, H. Aoi, T. Takeichi, T. Tsuruta, Chem. Lett., 645 (1976).
94. S. Yamashita, N. Yamawaki, H. Tani, Macromolecules, 6, 724 (1974).

95. R. J. P. Corriu, J. J. E. Moreau, J. Organometal. Chem., 120, 337 (1976).

96. K. Tamao, A. Minato, N. Miyake, T. Matsuda, Y. Kiso, M. Kumada, Chem. Lett., 133 (1975).

97. J. Halpern, Adv. Chem. Ser., 70, 1 (1968).

98. H. Brunner, Angew. Chem., 83, 274 (1971).

99. H. Brunner, Ann. N. Y. Acad. Sci., 239, 213 (1974).

100. H. Brunner, Topics in Current Chemistry, 56, 67 (1975).

101. K. Stanley, M. C. Baird, J. Amer. Chem. Soc., 97, 6598 (1975).

102. H. Brunner, J. Wachter, Chem. Ber., 110, 721 (1977).

103. W. S. Knowles, M. J. Sabacky, B. D. Vineyard, J. C. S. Chem. Comm., 10 (1972).

104. H. B. Kagan, T. P. Dang, J. Amer. Chem. Soc., 94, 6429 (1972).

105. A. C. Cope, C. R. Ganellin, H. W. Johnson, Jr., T. V. Van Auken, H. J. S. Winkles, J. Amer. Chem. Soc., 85, 3276 (1963).

106. For a review see: G. Paiaro, Organometal. Chem. Rev. A, 6, 319 (1970).

107. R. Lazzaroni, P. Salvadori, P. Pino, Chem. Comm., 1164 (1970).

108. Y. Kiso, K. Yamamoto, K. Tamao, M. Kumada, J. Amer. Chem. Soc., 94, 4373 (1972).

109. M. Idai, H. Ishiwatari, H. Yagi, E. Tanaka, K. Onozawa, Y. Uchida, J. C. S. Chem. Comm., 170 (1975).

110. Y. Ohgo, S. Takeuchi, Y. Natori, J. Yoshimura, Chem. Lett., 33 (1974).

111. J. Furukawa, S. Akutsu, T. Saegusa, Makromol. Chem., 81, 100 (1965).

112. T. P. Dang, J. C. Pouling, H. B. Kagan, J. Organometal. Chem., 91, 105 (1975).

113. W. Dumont, J. C. Poulin, T. P. Dang, H. B. Kagan, J. Amer. Chem. Soc., 95, 8295 (1973).

114. D. Y. Curtin, Record Chem. Progress, 15, 111 (1954).

115. G. Consiglio, Helv. Chim. Acta, 59, 124 (1976).

116. G. Gelbard, H. B. Kagan, R. Stern, Tetrahedron, 32, 233 (1976).

117. R. Glaser, Tetrahedron Lett., 2127 (1975).

118. G. Consiglio, P. Pino, in preparation.

119. S. Brunie, J. Mazan, N. Langlois, H. B. Kagan, J. Organometal. Chem., 114, 225 (1976).

120. W. S. Knowles, M. J. Sabacky, B. D. Vineyard, Adv. Chem. Ser., 132, 274 (1974).

PHOTOREDOX CATALYSIS

Thomas J. Meyer

The University of North Carolina

Chapel Hill, North Carolina 27514

The absorption of light at the molecular level leads to excited states having chemical and physical properties which are predictable and can be manipulated. It should be possible to exploit excited states or thermal intermediates which come from excited states in problems in homogeneous catalysis.

There are two different applications. In "photo-assisted catalysis", a reactive intermediate generated photochemically catalyzes a spontaneous (exoergic) reaction: oxygen activation, isomerization, polymerization, hydrogenation. Advantages of photochemical over thermal activation can be that: activation is localized at the light-acceptor site rather than by heating the whole medium; intermediates can be made which are inaccessible thermally; catalytic sites can be generated in unusual media (strong acids, hydrophobic regions) or at surfaces or interfaces (monolayers, micelles). Working catalytic systems could be based on known photochemical reactions including lig dissociation leading to coordinative unsaturation, isomerization to a reactive isomer, photochemical changes in a bound ligand, homolytic bond breaking (for example of a metal-metal band) to give radicals, changes in electron transfer, or photoejection of electrons. Possible applications of photo-assisted catalysis were discussed in some detail in a previous report within the more general context of energy conversion processes.[1]

In a second application, a chemical reaction is driven in the nonspontaneous or endoergic direction using light as a stoichiometric reagent. Here a light-driven chemical synthesis or "chemical photosynthesis" is involved. The associated catalytic system must absorb light, convert it into chemical energy, and catalyze the net reaction. The value of the application is, of course, that the

reaction driven could be the photoproduction of a fuel or other high
energy chemical and the external light source could be the sun.

The intention here is to concentrate on chemical photosynthesis,
and more narrowly on the use of metal complex excited states to drive
oxidation-reduction reaction catalytically. Such a narrow review
appears to be justified in the light of recent developments in this
area.

Electron Transfer Reactions of Excited States

Many important chemical transformations including the activation of hydrocarbons, the reduction of nitrogen or of carbon dioxide, and the photochemical splitting of water are net oxidation-
reduction reactions. Essentially all molecular electronic excited
states are potential redox reagents, because the absorption of light
leads to excitation of an electron to a higher level where it is
more weakly bound, and simultaneously, to an electron hole in a
lower level. In principle, it should be possible to bring this
excited state property to bear on the transformations mentioned
above. The problems involved are not trivial, but there is the
reassuring knowledge that such chemistry occurs around us constantly
through photosynthesis in biological systems.

Transition metal complexes offer several potential advantages
as photoredox catalysts:

1. In the proper ligand environment they can absorb light
strongly in the visible and near ultraviolet.

2. High chemical stability can be achieved.

3. Relatively fecile, chemically reversible redox processes
are known to occur for ground state complexes.

4. Metal ligand binding sites are available which can promote
reactions of coordinated ligands.

5. The nature of the lowest-lying excited state (metal, ligand,
or metal-ligand based) and its properties can be varied by systematic
synthetic changes.

The complex $Ru(bpy)_3^{2+}$ (bpy is 2,2'-bipyridine) absorbs light
strongly in the visible (λ_{max}450 nm, $\varepsilon= 1.4\times10^4$ in water) to give
a relatively long-lived excited state ($Ru(bpy)_3^{2+*}$, τ_0 ~600 μsec)[2,3]
which emits light at room temperature in a solution with a relatively
high efficiency.[4] One-electron reduction occurs by the addition of
an electron to the lowest unfilled level which is π^*(bpy) giving
$Ru(bpy)_3^+$ which is a strong reductant.[6] The results of spectroscopic and photochemical studies have revealed the following about
the $Ru(bpy)_3^{2+/2+*}$ system:

1. Absorption of light throughout the high energy visible and
near-ultraviolet gives the excited state $Ru(bpy)_3^{2+*}$ with high
efficiency.[7-9]

2. The emitting excited state is a Ru → bpy charge transfer
(CT) state in which considerable charge transfer has occured,

$$Ru^{II}(bpy)_3{}^{2+}(d^6) \xrightarrow{\;h\nu\;} Ru^{III}[(bpy)_3{}^{\bar{\cdot}}]^{2+*}(d^5\pi^*).^{8,9}$$

3. $Ru(bpy)_3{}^{2+}$ is stable to photolysis in water for extended periods.[10]

Electron transfer quenching of $Ru(bpy)_3{}^{2+*}$ by a series of ammine complexes of cobalt(III) was first reported by Gafney and Adamson (reaction 1)[2,11], but it was later suggested that their results could be explained by energy transfer to low-lying CT states of the Co(III) complexes which are known to give Co^{2+} with high efficiencies.[12] $Ru(bpy)_3{}^{2+*}$ can act as sensitizer

$$5H^+ + Ru(bpy)_3{}^{2+*} + Co(NH_3)_5Br^{2+} \longrightarrow Ru(bpy)_3{}^{3+} + Co^{2+} +$$
$$+ Br^- + NH_4{}^+ \tag{1}$$

or electron transfer reagent,[13] and, in fact, is known to undergo both energy and electron transfer processes with the same acceptor.[3,14]

Net electron transfer quenching is readily shown by flash photolysis.[3,15] At relatively high quencher concentrations and low concentrations of $Ru(bpy)_3{}^{2+}$, excitation and quenching occur during the flash and redox products are formed in small but spectrally observable amounts (reactions 1-3, in acetonitrile).[14]

Scheme I

$$Ru(bpy)_3{}^{2+} \xrightarrow{\;h\nu\;} Ru(bpy)_3{}^{2+*} \tag{1}$$

$$Ru(bpy)_3{}^{2+*} + PQ^{2+} \xrightarrow{\;k_q'\;} Ru(bpy)_3{}^{3+} + PQ^+ \tag{2}$$

$$Ru(bpy)_3{}^{3+} + PQ^+ \xrightarrow{\;k_b\;} Ru(bpy)_3{}^{2+} + PQ^{2+}$$

$$(PQ^{2+} \text{ is paraquat, } Me-N\hexagon\hexagon N-Me^{2+}) \tag{3}$$

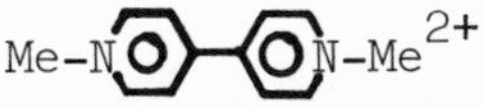

The total bimolecular quenching rate constant, electron and energy transfer, can be obtained by luminescence quenching experiments and k_b can be obtained directly by flash photolysis.[3,15] The excited state is relatively long-lived and if the quenching rate constant k_q is near the diffusion-controlled limit quencher concentrations of a few millimolar suffice to capture it completely by quenching.

The CT excited state has both oxidizing (Ru(III)) and reducing $[(bpy)_3^{\bar{\cdot}}]$ sites and in addition to oxidative quenching (reaction 2), can undergo reductive quenching[16] as shown by flash photolysis[17] (reactions 4-6, in acetonitrile).

Scheme II[17b]
$$Ru(bpy)_3{}^{2+} \xrightarrow{\;h\nu\;} Ru(bpy)_3{}^{2+*} \tag{4}$$

$$\text{Ru(bpy)}_3^{2+*} + \text{DMA} \xrightarrow{\quad k_q' \quad} \text{Ru(bpy)}_3^+ + \text{DMA}^+ \qquad (5)$$

$$\text{Ru(bpy)}_3^+ + \text{DMA}^+ \xrightarrow{\quad k_b \quad} \text{Ru(bpy)}_3^{2+} + \text{DMA} \qquad (6)$$

(DMA is N, N-dimethylaniline)

Ru(bpy)_3^+ has also been prepared in acetonitrile by electrochemical reduction of Ru(bpy)_3^{2+}.[17b]

An overall kinetic scheme for the quenching step (reactions 2 and 5) is given in Scheme III using the quenching of Ru(bpy)_3^{2+} by $\text{Fe(H}_2\text{O)}_6^{3+}$ in water as the example. Capture of the excited state by an external quencher

Scheme III[18].

$$\text{Ru(bpy)}_3^{2+*} + \text{Fe}^{3+} \rightleftharpoons \text{Ru(bpy)}_3^{2+*}, \text{Fe}^{3+} \xrightarrow{k_1} \text{Ru(bpy)}_3^{3+}, \text{Fe}^{2+}$$

$$1/\tau_0 \downarrow \uparrow h\nu \qquad \qquad \downarrow k_2 \qquad k_4 \qquad \qquad \downarrow k_3$$

$$\text{Ru(bpy)}_3^{2+} + \text{Fe}^{3+} \qquad \text{Ru(bpy)}_3^{2+}, \text{Fe}^{3+} \qquad \text{Ru(bpy)}_3^{3+} + \text{Fe}^{2+}$$

is in competition with excited state decay by radiative (k_{em}) and non-radiative pathways ($1/\tau_0 = k_{em} + k_{rd}$). Bimolecular quenching can occur by energy transfer or collisional deactivation (k_2), or by electron transfer (k_1). The appearance of the separated redox products Fe^{2+} and Ru(bpy)_3^{3+} marks a potentially exploitable light to chemical energy conversion. The competing pathways in the Scheme ultimately lead to heat. The efficiency of the conversion step depends on energy (k_2) vs. electron (k_1) transfer quenching and on the relative rates of separation (k_3) and back electron transfer (k_4) following electron transfer quenching. The electron transfer steps (k_1 and k_4) can be understood and rates predicted using the Marcus-Hush theories[19,20] for outer-sphere electron transfer.[3,15] The rate of the separation step depends on ion charge and on the dielectric and viscosity properties of the medium.[21]

Scheme III shows that several pathways compete with the formation of separated redox products. Yet at high quencher concentrations where essentially all excited states are captured by quenching, Lin and Sutin find that the efficiency of formation Fe^{2+} and Ru(bpy)_3^{3+} is 0.81 ± 0.16 at 25°.[10]

Redox potentials for Ru(bpy)_3^{2+*} acting as an electron transfer reagent have been estimated. The estimate was made by measuring quenching rates for a series of quenchers having slightly different oxidizing strengths.[22] The results are given in Table I as reduction potentials; values for the related ground state couples are also given there.

Using the data, the energetics of the steps in the electron transfer quenching schemes (I and II) can be calculated. Scheme I involves two conversion steps. Light energy is absorbed and converted into

Table I

Formal Reduction Potentials at Room Temperature (I = 0.1 $\underline{M}$)[16a, 22]

Couple	$\mathcal{E}'$,v(CH$_3$CN vs. SCE)	$\mathcal{E}'$,v(H$_2$O vs. NHE)
RuB$_3^{3+}$ + e → RuB$_3^{2+}$	1.29	1.26
RuB$_3^{2+*}$ + e → RuB$_3^{+}$	0.78	0.84
RuB$_3^{3+}$ + e → RuB$_3^{2+*}$	-0.81	-0.84
RuB$_3^{2+}$ + e → RuB$_3^{+}$	-1.32	-1.26

[a]In 1.0$\underline{M}$ H$_2$SO$_4$

molecular excited state energy in Ru(bpy)$_3^{2+*}$ (reaction 1). The
thermally equilibrated excited state is 2.10v above the ground state.
Light energy in excess of 2.10 v (590 nm), is absorbed to give
upper excited states or levels high in the vibrational manifold of
the lowest excited state. The excess energy is ultimately given off
as heat to the surroundings. In reaction 2 excited state energy of
2.10 v is converted into 1.7 v of chemical energy and 0.4 v of
heat (the potential for the PQ$^{2+/+}$ couple is -0.41 v in acetonitrile).
In the absence of a collection device, the chemical energy is only
transiently stored and recombination (reaction 3) returns the system
to equilibrium with the release of 1.7 v of heat.
     The ellect of excitation on the redox properties of Ru(bpy)$_3^{2+}$
is shown in the reduction potential diagram in Scheme IV.

Scheme IV

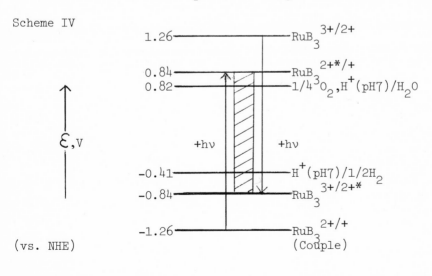

$\mathcal{E}$,V

(vs. NHE)

1.26——————————RuB$_3^{3+/2+}$

0.84—————————RuB$_3^{2+*/+}$
0.82—————————1/4³O$_2$,H$^+$(pH7)/H$_2$O

+hν   +hν

-0.41—————————H$^+$(pH7)/1/2H$_2$
-0.84—————————RuB$_3^{3+/2+*}$

-1.26—————————RuB$_3^{2+/+}$
                      (Couple)

$Ru(bpy)_3^{2+}$ is both a weak oxidant and reductant and is stable with respect to disproportionation (eq.7). The absorption of light enhances both oxidizing and reducing properties by the excited state energy (2.1V), and $Ru(bpy)_3^{2+*}$ is unstable with respect to disproportionation (eq. 8). The effect of the excitation is shown in Scheme IV by the light induced crossing of the redox energies for Ru(II) acting as reducant and oxidant.

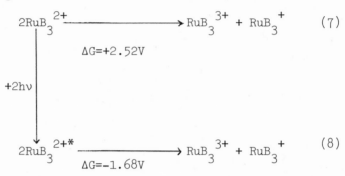

$$2RuB_3^{2+} \longrightarrow RuB_3^{3+} + RuB_3^+ \qquad (7)$$

$$\Delta G = +2.52V$$

$$+2h\nu$$

$$2RuB_3^{2+*} \longrightarrow RuB_3^{3+} + RuB_3^+ \qquad (8)$$

$$\Delta G = -1.68V$$

In Scheme IV the hatched area shows the energy difference between RuB_3^{2+*} as oxidant and reductant which is the disproportionation energy gap. $Ru(bpy)_3^{2+*}$ is thermodynamically capable of both oxidizing and reducing water at pH7 (eqs. 9 and 10) because the oxidation and reduction half cell potentials for water fall within the disproportionation energy gap. The

$$Ru(bpy)_3^{2+*} + 1/2H_2O \searrow \qquad \Delta G = -0.02V$$
$$Ru(bpy)_3^+ + 1/4O_2(g) + H^+(pH\ 7) \qquad (9)$$

$$Ru(bpy)_3^{2+*} + H^+(pH\ 7) \searrow \qquad \Delta G = -0.43V$$
$$Ru(bpy)_3^{3+} + 1/2H_2(g) \qquad (10)$$

water splitting energy is fixed at 1.23V (for $1/2H_2O \rightarrow 1/2H_2(g) + 1/4O_2(g)$) but the water half cell potentials vary with pH ($\mathcal{E}°$ ($O_2,H^+/H_2$) from 1.23V at pH 0 to 0.40V at pH 14). This means that pH variations can be used to adjust water redox potentials within an excited state redox gap or to bring them within access of another excited state.

Thermodynamically, all that is required for water splitting by redox steps is that the water half reactions fall within the energy difference between the ground and excited states (2.10V for $Ru(bpy)_3^{2+*}$). Ways of accomplishing this mechanistically include using the excited state as reductant and $Ru(bpy)_3^{3+}$ as oxidant (Scheme V), using the excited state as oxidant and $Ru(bpy)_3^+$ as reductant (Scheme VI), or by initial disproportionation (Scheme VII). The nergy gaps utilized in the three schemes are shown by the downward and upward pointing excitation arrows in Scheme IV and by the difference in the $Ru(bpy)_3^{3+/2+}$ and $Ru(bpy)_3^{2+/+}$ potentials (2.52V) where initial disproportionation is involved. Schemes like V, VI. and VII may, in fact, give some insight into how monolayer derivatives of $Ru(bpy)_3^{2+}$ operate when they photocatalyze the splitting of water.

Scheme V

	$\Delta G, v(pH7)$
$RuB_3^{2+} \xrightarrow{h\nu} RuB_3^{2+*}$	(+2.10)
$RuB_3^{2+*} + H^+ \longrightarrow RuB_3^{3+} + 1/2H_2(g)$	(−0.43)
$RuB_3^{3+} + 1/2H_2O \longrightarrow RuB_3^{2+} + 1/4O_2(g) + H^+$	(−0.44)
$1/2H_2O \xrightarrow{h\nu} 1/2H_2(g) + 1/4O_2(g)$	(+1.23)

$(\lambda \leqslant 590 \text{ nm } (2.10v))$

Scheme VI

$RuB_3^{2+} \xrightarrow{h\nu} RuB_3^{2+*}$	(+2.10)
$RuB_3^{2+*} + 1/2H_2O \longrightarrow RuB_3^+ + H^+ + 1/4O_2(g)$	(−0.02)
$RuB_3^+ + H^+ \longrightarrow 1/2H_2(g) + RuB_3^{2+}$	(−0.85)
$1/2H_2O \xrightarrow{h\nu} 1/2H_2(g) + 1/4O_2(g)$	(+1.23)

$(\lambda \leqslant 590 \text{ nm})$

Scheme VII

$2RuB_3^{2+} \xrightarrow{2h\nu} RuB_3^{2+*}$	(+4.20)
$2RuB_3^{2+*} \longrightarrow RuB_3^{3+} + RuB_3^+$	(−1.68)
$RuB_3^{3+} + 1/2H_2O \longrightarrow RuB_3^{2+} + 1/4O_2(g) + H^+$	(−0.44)
$RuB_3^+ + H^+ \longrightarrow RuB_3^{2+} + 1/2H_2(g)$	(−0.85)
$1/2H_2O \xrightarrow{2h\nu} 1/2H_2(g) + 1/4O_2(g)$	(+1.23)

$(\lambda \leqslant 590 \text{ nm})$

The discussion so far has been based on thermodynamic possibilities but some comments on mechanism are also appropriate. The redox chemistry of water is complicated because transformations between stable forms ($H_2O \rightarrow O_2$) involve multi-electron steps. Even if the excited state itself is incapible of reacting with water at a significant rate, it may be capable by simple electron transfer of driving coupled redox systems which can. When translated to the microscopic level, each of the net reactions in Schemes V, VI, and VII must be multiplied by at least 4 since any working mechanism must lead to integral numbers of H_2 and O_2 molecules produced. The disproportionation mechanism in Scheme VII is less efficient in terms of numbers of photons used but has the potential advantage kinetically of utilizing the stronger redox reagents $Ru(bpy)_3^{3+}$ and $Ru(bpy)_3^{+}$ and of increasing the redox energy gap to $\approx 2.52V$.

The electron transfer properties of $Ru(bpy)_3^{2+*}$ have been utilized in ways which lend support to possible photoredox applications. For example, it has been demonstrated by flash photolysis that $Ru(bpy)_3^{2+*}$ can be used as a photocatalyst for driving reactions transiently in the non spontaneous direction (Schemes VIII[24] and IX[17b], in acetonitrile). The reactions are driven using visible light even though neither of the reactants absorbs the light in the visible. The reason is that the energy demands on the reaction are independent of the optical properties of the reactants and depend instead on the free energy charge of the reaction which must fall within the excited state-ground state energy gap.

Scheme VIII[24]

$$Ru(bpy)_3^{2+} \xrightarrow{\ h\nu\ } Ru(bpy)_3^{2+*}$$

$$Ru(bpy)_3^{2+*} + PQ^{2+} \longrightarrow Ru(bpy)_3^{3+} + PQ^{+}$$

$$Ru(bpy)_3^{3+} + NPh_3 \longrightarrow Ru(bpy)_3^{2+} + NPh_3^{+}$$

$$NPh_3 + PQ^{2+} \xrightarrow[\Delta\mathcal{E}=1.45V]{\ h\nu\ } NPh_3^{+} + PQ^{+}$$

Scheme IX[17b]

$$Ru(bpy)_3^{2+} \xrightarrow{\ h\nu\ } Ru(bpy)_3^{2+*}$$

$$Ru(bpy)_3^{2+*} + DMA \longrightarrow Ru(bpy)_3^{+} + DMA^{+}$$

$$Ru(bpy)_3^{+} + O_2 \longrightarrow Ru(bpy)_3^{2+} + O_2^{-}$$

$$DMA + O_2 \longrightarrow DMA^{+} + O_2^{-}$$

The chemistry in Scheme IX may be important since the super-oxide ion is an "activated" form of oxygen. If the combined ox-idizing strengths of O_2^- (or HO_2) and the quencher can be brought to bear on appropriate substances, it may be possible to develop useful photocatalyzed oxidation or initiation steps involving O_2. Noteworthy in this context is the work of Demas et al. on the photocatalytic production of singlet oxygen using a series of metal complex excited states[25] and of Hammond et al. on the production of H_2O in acidic solutions containing O_2 and $Ru(bpy)_3^{2+*}$ can be simultaneously oxidized and reduced in the same solution when both oxidizing and reducing quenchers are present. Taken together the two reactions lead to the chemically catalyzed disproportionation of $Ru(bpy)_3^{2+*}$ (eq. 8) shown in Scheme X.[27]

Scheme X[27]

$$2RuB_3^{2+} \xrightarrow{2h\nu} 2RuB_3^{2+*}$$

$$RuB_3^{2+*} + TMPD \longrightarrow RuB_3^+ + TMPD^+$$

$$RuB_3^{2+*} + PQ^{2+} \longrightarrow RuB_3^{3+} + PQ^+$$

$$2RuB_3^{2+} + PQ^{2+} + TMPD \xrightarrow{2h\nu} RuB_3^{3+} + RuB_3^+ + PQ^+ + TMPD^+$$

The work described so far has relied largely on $Ru(bpy)_3^{2+*}$, but electron transfer appears to be a common reaction for metal complex excited states. It has been shown that electron transfer quenching can occur for a series of CT states,[3,15,28] and for f-f and π-π^* excied states.[28] Even more striking is the use of flash photolysis to show that at high quencher concentrations even short-lived excited states which are not observable spectrally using conventional techniques can be captured by electron transfer quench-ing (e.g., reaction 9, in DMF).[29] The use of high quencher

$$Ru(TPP)(py)_2^* + Ru(NH_3)_6^{3+} \longrightarrow Ru(TPP)(py)_2^+ + Ru(NH_3)_6^{2+}$$

(TPP is tetraphenylporphine)   (9)

concentrations or intramolecular quencher sites may allow for el-ectron transfer quenching of very short lived excited states, perhaps even of upper excited states of the light absorber, and may expand immensely the number of excited states available for photo-redox applications.

Limitations and Strategies

The photochemical approach may lead to useful applications in the catalysis of spontaneous reactions, and in the stoichiometric photoproduction of chemicals using visible light. Severe limitations

can be imposed by competitive degradation processes. Even low yield
photodegradation can limit the operation of a catalytic system to
an unacceptably short period. In practical systems it may be
necessary to use simple aquo ions, unusually stable metal complexes,
or protective chemical environments or matrices. Any large scale
chemical photosynthetic application will, of course, demand perhaps
unacceptably large surface areas because of the low intensity of the
solar insolation.

In considering possible photoredox applications, the splitting
of water is a useful example. From the photochemical point of view,
the critical factors involved have been alluded to in the previous
section. They are summarized in a generalized form in· Scheme XI.
The photochemical part of any working system will usually consist of
three parts: light absorption, energy conversion, and interfering
recombination steps. The emphasis here is on redox processes. It
should be noted that the photocatalytic splitting of water has also
been discussed based on photochemical cycles which involve metal[29]
hydrides and binuclear metal complexes.

Scheme XI

$$D \xrightarrow{\hspace{0.3em}h\nu\hspace{0.3em}} D^* \qquad\qquad\qquad \text{absorption}$$

$$D^* + Q \longrightarrow D^+ + Q^-$$

$$D^* + Q' \longrightarrow D^- + Q'^+$$

$$D^* + H^+ \longrightarrow D^+ + 1/2H_2 \qquad\qquad \text{conversion}$$

$$D^* + H_2O \longrightarrow D^- + H^+ + 1/4O_2$$

$$2D^* \longrightarrow D^+ + D^-$$

$$D^+ + D^- \longrightarrow D + Q$$

$$D^- + Q'^+ \longrightarrow D + Q' \qquad\qquad \text{recombination}$$

The necessary first step in any scheme is light absorption.
About one-third of the solar insolation at the earth's surface
occurs from the near uv to 610 nm. Long wavelength light may not
be usable for molecular systems since low-lying excited states tend
to undergo rapid conversion processes to the ground state. In
order to minimize material needs, high absorptivity is necessary
in the visible which probably demands complex ligand systems such
as the porphyrins. Absorption bands have finite band widths and
to cover a broad region of the spectrum may require building

in two or more overlapping absorption bands. If so, conversion
processes from from upper excited states to the excited state used
in catalysis must be efficient. As mentioned in the previous section, energy limitations in the system are fixed by the energy of
the thermally equilibrated excited state. With light of lower
energy, the excited state can not be reached; the excess energy
of shorter wavelength light ultimately appears as heat. Light
absorption can be separated from later steps since excited state
energy can be passed on by energy transfer to a second site where
activation-conversion may occur.

The light to chemical energy conversion step could occur by
direct reactions of the excited state acting as oxidant or reductant or by initial electron transfer quenching. In an electron
transfer quenching mechanism, separated redox reagents are created.
The redox intermediates may carry out the desired chemistry themselves or may be used to drive coupled redox systems which do. In
either case recombination reactions represent a major problem. For
example, in Schemes I and II, the recombination reactions 3 and 6
occur on the msec time scale under the conditions of the experiments. Any collection device for the chemical energy produced must
operate on this or a shorter time scale. Recombination can be
overcom if the reactions driven are rapid and irreversible. or if
systems are designed in which back electron transfer is relatively
slow. One approach is to design interfacial and membrane systems
where the oxidizing and reducing sites are created and transported
rapidly into different phases or onto opposite sides of a membrane.
Other possibilities include semiconductor interfaces where the
unidirectional nature of electron transfer arising from band bending
at the surface can be exploited and molecular systems where the
special properties of excited states can be used to obtain "molecular rectification".

Using chemistry already discussed, it is possible to design
schemes which, in principle, could lead to the photocatalytic
splitting of water. For example, in the series of reactions in
Scheme XII the disproportionation of $Ru(bpy)_3^{2+*}$ into $Ru(bpy)_3^{3+}$
and $Ru(bpy)_3^{+}$ is catalyzed by the presence of both oxidizing and
reducing quenchers. $Ru(bpy)_3^{3+}$ is thermodynamically capable

Scheme XII

$$2RuB_3^{2+} \xrightarrow{\quad 2h\nu \quad} 2RuB_3^{2+*}$$

$$RuB_3^{2+*} + PQ^{2+} \longrightarrow RuB_3^{3+} + PQ^{+}$$

$$RuB_3^{2+*} + DMA \longrightarrow RuB_3^{+} + DMA^{+}$$

$$RuB_3^{3+} + 1/2H_2O \longrightarrow RuB_3^{2+} + 1/4O_2 + H^{+}$$

$$RuB_3^{3+} + H^{+} \longrightarrow 1/2H_2(g) + RuB_3^{2+}$$

Scheme XII (continued)

$$PQ^+ + DMA^+ \longrightarrow PQ^{2+} + DMA$$

$$1/2H_2O \xrightarrow{2h\nu} 1/2H_2(g) + 1/4O_2(g)$$

of oxidizing water to oxygen and $Ru(bpy)_3^+$ of reducing it to hydrogen.

However, the scheme also points out that in addition to the photochemical framework in Scheme XI, there is a second part to the problem namely the collection of the chemical energy once produced. In the case of water splitting, collection occurs by oxidation and reduction of water and a major difficulty arises because of the multi-electron nature of the oxidation and reduction processes. $Ru(bpy)_3^{3+}$ does oxidize water, but not at a great rate at pH 7, and $Ru(bpy)_3^{2+}$ is not produced cleanly apparently because of the indiscriminate nature of intermediates produced by one-electron oxidation steps.[31] The problem is illustrated further in reactions 10 and 11. With either $Ru(bpy)_3^{2+*}$ or

$$RuB_3^{2+*} + H^+(pH7) \xrightarrow[\Delta G=+1.67 \text{ v}]{} RuB_3^{3+} + H(g) \qquad (10)$$

$$RuB_3^+ + H^+(pH7) \xrightarrow[\Delta G=+1.29 \text{ v}]{} RuB_3^{2+} + H(g) \qquad (11)$$

$Ru(bpy)_3^+$ as reductants, simple one-electron transfer to H^+ to give H is highly unfavorable thermodynamically. One-electron transfer steps are involved in the production of H_2 or O_2 by ultraviolet photolysis of aquo ions in water.[32] In fact, both are produced by UV photolysis of solutions containing Ce(III) and Ce(IV).[33] However, the energy requirements are such that shorter wavelength UV light must be used and quantum efficiencies are low because of recombination and secondary free radical-type reactions.

There are a number of possible approaches to photoredox catalysis which may avoid simple one-electron transfer steps. Generation of strongly reducing $Ru(bpy)_3^+$ might be coulped to a secondary system for which hydrogen evolution is facile

$$Ru(bpy)_3^+ + M \longrightarrow Ru(bpy)_3^{2+} + M^-$$

$$M^- + H^+ \longrightarrow MH \xrightarrow{+h\nu} M + 1/2H_2$$

$$M^- + H_2O \longrightarrow M + 1/2H_2(g) + OH^-$$

or it might be possible to create a chemical environment where the one-electron transfer product is highly stabilized as in the case

of chemisorption of hydrogen in metals. Another approach is to
devise systems which like chloroplasts can absorb and utilize more
than one photon simultaneously. In a properly designed, chemically-
linked system in which there are a series of light acceptor sites,
energy transfer between sires could lead to adjacent states which
undergo simultaneous electron transfer to give H_2 directly.

$$D^* \quad D^* \quad + \quad 2H_2O \qquad H_2(g) + 2OH^- + \quad D^+ \quad D^+$$

The use of more than one photon could also lead to an upper or
doubly excited state localized at a single site. An appropriate
site might be capable of simultaneous two-electron transfer, or it
might be sufficiently energetic to produce highly unstable one-
electron intermediates.

References

1. National Science Foundation Conference on "Inorganic Photochemistry Related to Transfer, Storage, Conservation, and Conversion
of Energy", Dulles Marriott Hotel, June 5-6, 1975.
2. G. Navon and N. Sutin, Inorg. Chem., 13, 2159 (1974).
3. C. R. Bock, T. J. Meyer, and D. G. Whitten, J. Am. Chem. Soc.,
96, 4710 (1974); R. C. Young, C. R. Bock, T. J. Meyer, and D. G.
Whitten, in preparation.
4. F. E. Lytle and D. M. Hercules, J. Am. Chem. Soc., 91, 253 (1969).
5. F. E. Lytle and D. M. Hercules, Photochem. Photobiol., 13, 123
(1971); A. M. Sargeson and D. A. Buckingham, in "Chelating Agents
and Metal Chelates", F. P. Dwyer and D. P. Mellor, Ed., Academic
Press, N.Y., 1964, Chapt. 6.
6. N. E. Tokel-Takvoryan, R. E. Hemingway, and A. J. Bard, J. Am.
Chem. Soc., 95, 6582 (1973); T. Saji and S. Aoyagui, J. Electroanal.
Chem. Interfacial Electrochem., 58, 401 (1975).
7. F. Bolletta, M. Maestri, and V. Balzani, J. Phys. Chem., 80,
2499 (1976).
8. K. W. Hipps and G. W. Crosby, J. Am. Chem. Soc., 97, 7042 (1975)
and references therein.
9. J. Van Houten and R. J. Watts, J. Am. Chem. Soc., 98, 4853 (1976).
10. C. T. Lin and N. Sutin, J. Phys. Chem., 80, 97 (1976); J. Am.
Chem. Soc., 97, 3543 (1975).
11. H. D. Gafney and A. W. Adamson, J. Am. Chem. Soc., 94, 8238
1972).
12. P. Natarajan and J. F. Endicott, J. Am. Chem. Soc., 94, 3635
(1973); J. Phys. Chem., 77, 1823 (1973).
13. V. Balzani, L. Mozzi, M. F. Manfrin, F. Bolletta, and G. S.
Laurence, Coord. Chem. Rev., 14, 321 (1975).
14. A. R. Gutierrez, T. J. Meyer, and D. G. Whitten, Molec. Photochem., 7, 349 (1976).

15. C. T. Lin, W. Boettcher, M. Chou, C. Creutz, and N. Sutin, J. Am. Chem. Soc., 98, 6536 (1976).

16. (a) C. Creutz and N. Sutin, Inorg. Chem., 15, 496 (1976); (b) A. Juris, M. T. Gandolfi, M. F. Manfrin, and V. Balzani, J. Am. Chem. Soc., 98, 1047 (1976).

17. (a) C. Creutz and N. Sutin, J. Am. Chem. Soc., 98, 6384 (1976). (b) C. P. Anderson, D. J. Salmon, R. C. Young, and T. J. Meyer, J. Am. Chem. Soc., in press.

18. D. Rehm and A. Weller, Ber. Bunsenges. Phys. Chem., 73, 834 (1969).

19. R. A. Marcus, J. Phys. Chem., 43, 679 (1975); Ann. Rev. Phys. Chem., 15, 155 (1964); N. S. Hush, Trans. Faraday Soc., 57, 557 (1961).

20. N. Sutin in "Inorganic Biochemistry", Vol. 2, G. L. Eichhorn, Ed., Elsevier, Scientific Publications Co., N.Y., 611 (1973).

21. R. C. Jarnagin, Accounts Chem. Res., 4, 420 (1971); M. Eigen, W. Kruse, G. Moass, and D. L. DeMaeyer, Prog. Reaction Kinetics, 2, 287 (1964).

22. C. R. Bock, T. J. Meyer, and D. G. Whitten, J. Am. Chem. Soc., 97, 2909 (1975).

23. G. Sprintschnik, H. W. Sprintschnik, P. P. Kirsch, and D. G. Whitten, J. Am. Chem. Soc., 98, 2337 (1976).

24. R. C. Young, T. J. Meyer, and D. G. Whitten, J. Am. Chem. Soc., 97, 16 (1975).

25. J. N. Demas, D. Diemente, and E. W. Harris, J. Am. Chem. Soc., 95, 6864 (1973); J. N. Demas, E. W. Harris, C. M. Flynn, Jr., and D. Diemente. J. Am. Chem. Soc., 97, 3838 (1975).

26. J. S. Winterle, D. S. Kliger, and G.S. Hammond, J. Am. Chem. Soc., 98, 3719 (1976).

27. J. Nagel, T. J. Meyer, and R. C. Young, manuscript in preparation.

28. R. C. Young, T. J. Meyer, and D. G. Whitten, J. Am. Chem. Soc., 98, 286 (1976).

29. R. C. Young, T. J. Meyer, and D. G. Whitten, submitted

30. V. Balzani, L. Moggi, M. F. Manfrin, F. Bolletta, and M. Gleria, Science, 189, 852 (1975).

31. C. Creutz and N. Sutin, Proc. Natl. Acad. Sci., U.S.A., 72, 2858 (1975).

32. M. D. Archer, Photochemistry, Vol. 6, Specialist Periodical Reports, The Chemical Society, 1975, Part V.

33. L. J. Heidt and A. F. McMillan, J. Am. Chem. Soc., 76, 2135 (1954).

I. HYDROCARBON CONVERSION

Co-Chairmen: L. Cassar, D. R. Coulson
Members: B. Calcagno, C. P. Casey, S. Cenini, M. Clerici,
          F. Conti, G. Deganello, D. R. Fahey, C. Masters,
          J. Osborn, R. R. Schrock, A. Yamamoto

The decreasing availability of hydrocarbon feedstocks demands
that they be utilized at maximum efficiency. Many current industrial
processes involving hydrocarbons are carried out using heterogeneous
catalysis. Replacement of these processes by ones utilizing homogeneous catalysis offers the potential of yielding both less energy-
intensive and more selective processes.

In the following we outline some desirable goals involving the
utilization of homogeneous catalysis in several areas of hydrocarbon
conversion. Closely related conversions involving functionalized
hydrocarbons have also been included for completeness.

A. Hydrogenation and Isomerization of Unsaturated Hydrocarbons

Olefin hydrogenations are important industrial processes. Often
the achievement of selective hydrogenations is critical to the success
of such processes.

There is a continuing need to search for and to study novel hydrogenation catalysts. Such complexes might contain metals outside
of Group VIII, new ligand systems, and/or more than one metal.
Through such basic studies we may hope to discover new and more selective hydrogenation catalysts. An example of the type of selectivity
desired is illustrated by the selective hydrogenation of an internal
double bond in the presence of one or more terminal double bonds.
Few catalysts are currently known which demonstrate such selectivity.

Another important goal involves the need for catalysts which are active in the presence of conventional "poisons" such as CO (a possible contaminant in H_2 obtained from synthesis gas) or sulfur-containing compounds (possible contaminants in feedstocks obtained from coal).

The hydrogenation of aromatic systems is likely to become increasingly important, especially if coal, which contains polyaromatic compounds, becomes an important hydrocarbon source for the chemical industry. Of special interest here are processes involving the selective hydrogenation of aromatic compounds. One such example is found in the hydrogenation of benzene to cyclohexene. Homogeneous catalysts could be valuable here in increasing the selectivity to cyclohexene.

In addition to catalysts for the hydrogenation of olefinic systems there also exists a need for catalysts which can hydrogenate functional groups such as C=O, C=N, $-NO_2$ and $-C\equiv N$, not easily reduced under the usually mild conditions of homogeneous catalysis. A valuable goal is to be able to specifically hydrogenate these functionalities in the presence of other sensitive functionalities, e.g.

The area of olefin isomerization contains many unsolved problems. For convenience we differentiate two basic types of olefinic isomerization: (1) cis-trans isomerization involving a change of relative substituent positions around a given olefinic bond and (2) hydrogen-transfer isomerizations involving the movement of an olefinic bond along a carbon skeleton. Currently, there exist too few homogeneous catalysts which are able to selectively effect type (1) isomerizations to the exclusion of type (2). Also there is a need for catalysts which can effect a type (2) isomerization on only certain olefinic bonds in a molecule while leaving others untouched.

It is well known that certain terminal olefins are often easily isomerized to the normally favored internal olefins. However, the less favored isomer may be utilized if (1) the rate of attaining equilibrium between the two is rapid and if (2) the small amount of terminal olefin formed reacts much more rapidly than the internal one in a reaction catalyzed by the same (or a different) catalyst. These coupled reactions would allow one to convert the mixture of internal olefins to one terminally functionalized derivative, e.g.

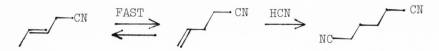

This could be an important concept in more efficiently utilizing olefin feedstocks.

B.   Polymerization, Metathesis and Oligomerization of Olefins

1.   Polymerization. Present technology involving the polymerization of ethylene and propylene using Ziegler-Natta catalysts has reached an advanced stage, and process technology continues to improve. Despite the advanced state of the art in this area, many important aspects of the mechanism of Ziegler-Natta processes still remain obscure. Heterogeneous gas-phase polymerizations are expected to dominate future developments and no clear need for homogeneous systems would seem to exist. However, mixtures of butenes are known to be more difficult to polymerize with Ziegler-Natta catalysts. Future prices may make this and related processes more attractive and prompt more investigations of such processes. Homogeneous catalysis might be used to effect more efficient and selective polymerization by, for example, rapid, prior isomerization of the but-2-ene or by inducing co-polymerization of the terminal and internal olefins. In addition, studies on well defined homogeneous systems would seem both necessary, from a fundamental viewpoint, and important if a new generation of more sophisticated catalysts are to be developed. With respect to polymerization of dienes, the factors which determine mode and stereochemistry of polymerization by coordination catalysts are not completely understood. This is especially true of those steric effects at the metal center which govern cis-1,4 vs trans-1,4 addition or 1,4 vs 1,2 addition.

Recent advances have shown that improved properties and better performances of polymers may be obtained by co-polymerization of different monoolefins to yield either the alternate or block type co-polymers. Present methods for producing such co-polymers are largely restricted to monomers subject to anionic polymerization. Also, many functional groups cannot be tolerated. Hence, blending of ordinarily incompatible homo-polymers by block co-polymerization, where the length of the sequence of the monomer domains are regulated, could have great practical utility. Metal catalysis could show promise in this area. For example, the use of catalysts which employ two metals in which rapid trans-alkylation is possible may enable separate chain growths for controlled periods. Such catalysts could be extremely important for co-polymerizations of monomers with highly dissimilar properties (e.g. ethylene plus acrylonitrile, propylene plus acrylate esters) as well as those with different reactive functions (e.g. the reactions of olefins with epoxides). Such ambitious goals may seem to be realizable only by the use of soluble, transition metal catalysts; however, other means of activation (e.g. chelating Li catalysts, etc.) could also be very effective.

2. Olefin Metathesis. Activity in this fascinating and important area of catalysis continues to be intense. Recent advances in the understanding of some details of the mechanisms provide a stimulus for further studies which may reveal the potential breadth of this type of catalytic reaction. It is clearly too early to assess either the potential utility of these reactions or the actual mechanisms involved. The area of olefin metathesis must be tackled on at least two fronts for its full potential to be realized. Firstly, an understanding of the precise details of the mechanism (or mechanisms) is imperative. The initiation process remains obscure, yet it could in itself provide a basis for whole new areas of catalysis based on carbene complexes if certain of the recently suggested initiation mechanisms are operative. Also, in this connection, preparation and reactivity studies on metal-carbene compounds would be helpful. The chain transfer process -- now better understood but lacking details -- must be further elucidated. Secondly the development of new, well defined metathesis catalysts is essential. This is necessary for several reasons. With these new catalysts one may expect (1) reaction systems permitting an easier delineation of the mechanism(s) (2) higher activities (3) lessened Lewis acid activities leading to higher selectivities and, perhaps most importantly, (4) catalysts active with functionalized olefins. While this latter objective has proven difficult to achieve, should it be realized, synthetic chemistry would be greatly benefited. Some examples of such transformations which could be industrially useful include

For cyclic olefins, the use of metathesis for polymerization, and in particular, the production of polypentenamer from cyclopentene is well known. However, co-polymers of functionalized cyclic olefins should show extremely interesting properties, e.g. good dyability, adhesion properties, etc. In addition, extension of metathesis catalysis to heterocyclic olefins would be most interesting since such polymer products are practically unknown (e.g. dehydrofuran polymers).

Work on metathesis of other unsaturated functions should be a further objective. For example, acetylene metathesis and even metathesis involving such groups as -N=N-, $>$C=N-, -C≡N, may be important.

3. Oligomerization. Since the process of olefin oligomerization is an integral phase of olefin polymerization, better knowledge of the discrete steps occurring during the former process should be relevant to the latter. For example, the problems of stereoselectivity and regioselectivity are common to both processes but may be more easily studied in oligomerizations. A particular problem in this area is concerned with the oligomerization of asymmetric dienes such as isoprene. Recent studies using soluble palladium catalysts have shown how a certain degree of control of coupling in the oligomerization can be achieved by careful modification of steric and electronic effects at the palladium atom. Detailed mechanistic studies on such systems should be undertaken. In addition, control of regioselectivity during the oligomerization of isoprene could be valuable towards controlling the formation of valuable natural products intermediates.

An area having a synthetic potential comparable to olefin metathesis is concerned with the oligomerization of functionalized olefins. Progress has been slow in this area; however, examples are known such as the ruthenium-catalyzed dimerization of acrylonitrile to yield 1,4-dicyanobutenes. The major problem here is associated with yields of less than ca. 60% of useful products.

Examples of co-oligomerizations of simple olefins are known. However, more information regarding these processes is desirable. Furthermore, co-oligomerization of functionalized olefins with simple dienes should be studied further. Important examples include the reported formation of C_{11} linear carboxylic esters from butadiene and acrylate esters as well as C_{10} linear dicarboxylic esters from these same reactants. The simultaneous introduction of functional groups (e.g. -OH, -NH$_2$, -CO$_2$R) during co-oligomerization is also a desirable synthetic goal. Very few examples of this sort of chemistry now exist.

A new approach to heterocyclic synthesis involves the co-oligomerization of molecules containing heteroatoms (e.g. R-CN, NH$_3$) with olefins, acetylenes or dienes. For example pyridine derivatives have been prepared by the Co-catalyzed co-oligomerization of acetylenes with nitriles. Certainly these and related concepts deserve more investigation since new synthetic methods are likely to be discovered.

C. Activation of Saturated Hydrocarbons

Reactions involving the breaking of C-H and/or C-C bonds are obviously important to the general area of hydrocarbon conversions. For the most part work involving such reactions has tended to concentrate on substrates containing functional groups which serve to activate or assist the desired transformation. In view of this, more work is needed on systems involving simple alkanes. To accomplish this there is a strong need to (1) develop an understanding of the basic chemistry involved in existing examples of alkane catalysis and (2) undertake an extensive search for new metal systems capable of acting as catalysts in alkane transformations. As an example of an approach that may be taken to construct new catalysts one may consider the following possible modes of C-H activation. A dissociative addition of a C-H bond to a metal site may be envisaged as a one- or two-site process, i.e.,

$$M \ + \ R\text{-}H \ \longrightarrow \ \overset{R}{\underset{M}{\diagdown}}\overset{H}{\diagup}$$

$$M_1\text{-}M_2 \ + \ R\text{-}H \ \longrightarrow \ \underset{M_1 \text{---} M_2}{\overset{R \qquad H}{| \qquad |}}$$

The latter mode of addition, which has parallels in heterogeneous catalysis, suggests the use of cluster complexes as catalysts. There may also be certain advantages to using bimetallic cluster complexes where the presence of two different metal centers are necessary to facilitate the breakage of C-H bonds.

Another possible approach is found in the well-known Pt^{+2} homogeneously catalyzed H/D exchange reaction of alkanes with D_2O. In this system multiple exchange has been demonstrated. This suggests the possible intermediacy of metal carbene species (M=CHR). Systems known to favor the formation of these species should be studied in the context of alkane activation.

In order to promote catalyst-substrate interactions, vacant sites are necessary on the metal. Methods of creating coordinatively unsaturated metal centers in hydrocarbon environments should be studied, e.g. photolysis of metal complexes leading to ligand dissociation (creation of vacant sites) and possibly promoting the metal to higher electronic states (photoassisted catalysis). The electrochemical generation of metals in unusual oxidation states generally unattainable by conventional means could also be used in this context.

One-electron transfer processes should also be considered in the context of C-H/C-C activation. Such processes are already well known, and are considered further in the section of this report dealing with the subject of selective oxidations.

Close parallels to the reactions discussed here already exist in both enzyme and heterogeneous catalysis. This suggests that fruit-ful comparisons may be made between these areas and metal-complex chemistry. For example, the use of metal clusters as catalysts has parallels in heterogeneous catalysis, as already pointed out. Anoth-er area which invites comparison to metal-complex catalysis involves the well-known reactions of alkanes with "super-acid" systems. These studies encourage the possibility of developing highly electrophilic metal complexes which could lead to catalytic activation of both C-C and C-H bond-containing systems.

Catalytic reforming, as practiced by the petrochemical industry, is a high temperature ($>400^{\circ}C$) process. Important improvements in selectivity could be achieved by developing more active homogeneous catalysts. Since current heterogeneous catalysts are bimetallic (e.g. Pt/Ir) the study of such systems in the homogeneous mode may lead to greater insight into the mechanism of the heterogeneous processes. Since the support, e.g. Al_2O_3, SiO_2 can have a dramatic effect on the activity of heterogeneous catalysts, it is also sug-gested that future work on homogeneous catalysts include the study of interactions of metal oxides with transition metal complexes.

An important consideration in all aspects of hydrocarbon conver-sions is the question of thermodynamic feasibility of the initial steps involving C-C and/or C-H bond cleavage with metal complexes. In fact, information currently available suggests that the formal oxidative addition of the C-H bond of alkanes to most metal complexes is thermodynamically disfavored as evidenced by the well-known ten-dency of M(H)R systems to collapse yielding the alkane R-H among the products. Nevertheless, little is currently known of the thermody-namics of such reactions and it seems likely that extended studies involving the determination of the heats of formation of transition metal alkyl, carbene, and hydride complexes should identify systems potentially capable of activating C-H and C-C bonds.

D. Transformation of Aromatic Hydrocarbons

Transformations of aromatic hydrocarbons are important indus-trial processes. Among these the reactions involving alkyl aromatics are preeminent. For example, the catalytic isomerization of xylenes is an important step in the production of p-xylene, an important fi-ber intermediate. Currently, this process is effected in good yields by heterogeneous catalysis at moderate temperatures ($150-250^{\circ}C$).

While homogeneous catalysis offers the possibility of operating at
lower temperature it is perhaps more significant that an increased
selectivity may also be achievable to avoid formation of ethyl benzene, a significant by-product in current processes. Similar con-
siderations apply to the dimethylnapthalenes which are emerging as
important fiber intermediates.

Related to these processes are those involving the de-alkylation
and trans-alkylation of alkylaromatics. These reactions are important in the synthesis of a wide variety of aromatic hydrocarbons.
Selectivity is of primary importance in these reactions. For example, finding a catalyst which could achieve the highly selective
formation of p-xylene from toluene via a trans-alkylation reaction
would be of immense value in reducing the cost of this intermediate.

Current heterogeneous catalyst systems lack the desired selectivity
in this reaction.

Many other reactions of aromatic hydrocarbons could be improved
through the intervention of homogeneous catalysis. The classic
isomer distribution problems associated with the usual electrophilic
substitution reactions could, in principle, be eliminated through
the introduction of new catalytic methods of functionalization and
isomerization. Currently, little has been done in this area but
existing evidence encourages the possibility that new patterns of
aromatic substitution are attainable through homogeneous catalysis.

In addition, the coupling of monocyclic aromatics could be an
important source of functional biphenyls. Recent work has shown
that this reaction is homogeneously catalyzed but problems in selectivity remain.

In related homogeneously catalyzed systems dramatic improvements in selectivity have been achieved by ligand modification. This suggests that similar improvements could be realized here.

An important feature of homogeneous systems is their ability to function efficiently under mild conditions. This could allow selective transformtaion of aromatic compounds containing sensitive functional groups.

The aromatization of alkanes and alkenes is an important route to aromatics. Future feedstocks (e.g. via the Fischer-Tropsch process) are likely to be low in aromatics necessitating their formation via conversion of the predominant alkane/alkene content of these feedstocks. Currently, aromatization is necessarily a high temperature process because of thermodynamic limitations. However, oxidative aromatization of hydrocarbons is, in principle, a favorable process at low temperatures. Homogeneous catalysis could be used to effect these transformations.

Recommendations

1. The study and preparation of more selective hydrogenation catalysts.
2. Extended studies of the mechanisms of olefin and diene polymerizations catalyzed by homogeneous metal systems.
3. The study of co-polymerization of radically different co-monomers as a means of obtaining novel block co-polymers.
4. Extensive studies of olefin metathesis reactions from both a mechanistic and a preparative point of view.
5. Studies concerned with the achievement of metathesis reactions of functionalized olefins.
6. Further studies of the factors which control selectivity in oligomerizations of simple olefins or functionalized olefins and dienes.
7. Studies on new methods of synthesis of heterocyclic compounds via transition metal catalyzed cyclooligomerizations.
8. An increased effort on the study of the mode of interaction of C-H/C-C bonds with metal centers with a concomitant search for new catalytic systems.
9. An extensive thermodynamic characterization of alkyl, hydride, and carbene containing metal complexes.
10. Increased effort on improving the selectivity of oxidative conversions of hydrocarbons.

Bibliography

Included below are a few general references appropriate to each of the topics discussed above.

A. Hydrogenation and Isomerization
    F. J. McQuillen, Homogeneous Hydrogenation in Organic Chemistry,
Dortrecht, Boston, 1976.
    B. R. James, Homogeneous Hydrogenation, Wiley, New York, 1973.
B. Polymerization, Metathesis and Oligomerization
    I. D. Rubin, Poly(1-Butene)-its Preparation and Properties,
Gordon and Breach, N.Y., 1975.
    A. D. Ketley, Ed., Stereochemistry of Macromolecules, Vol. I,
Arnold, 1967.
    P. N. Rylander, Organic Synthesis with Noble Metal Catalysts,
Academic Press, N.Y., 1973.
    N. Calderon, E. A. Ofstead and W. A. Judy, Ang. Chem., Int. Ed.,
15, No. 7, 401 (1976).
C. Activation of Saturated Hydrocarbons
    N. F. Goldshleger, M. L. Khidekel, A. E. Shilov and A. A.
Shleinman, Kinet. Katal., 1974, 15, 261.
    R. J. Hodges, D. E. Webster and P. B. Wells, J. Chem. Soc.,A,
1971, 3230.
D. Transformations of Aromatic Hydrocarbons
    G. W. Parshall, Accts. Chem. Res., 8, 113 (1975).
    F. R. S. Clark, R. O. C. Norman and C. B. Thomas, J. Chem. Soc.,
A, 1975, 121.
    H. Itatani and H. Yoshimoto, J. Org. Chem. 1973, 38, 76.

II. HOMOGENEOUS SELECTIVE OXIDATION

Co-Chairman: G. Modena, K. B. Sharpless
Members: G. Costa, J. Halpern, Y. Ishii, B. R. James,
J. Lyons, H. Mimoun, P. Rossi, B. Sheldon,
P. Teyessie

The search for better (i.e., more selective) oxidizing agents
has been going on for many years in both academic and industrial
laboratories. Although there have been successes, some of the key
problems have not been solved. For obvious reasons one would like
to use molecular oxygen as the oxidizing species whenever possible.
Most conceivable oxidations of organic compounds are highly exothermic but kinetically slow due in part to the triplet nature of the
ground state oxygen. This means that oxidations are hard to start,
but once underway, they are difficult to halt short of carbon dioxide
and water. Thus, the overwhelming concern in oxidation chemistry is
selectivity. It is our purpose to try and identify the ways in which
homogeneous catalysis might aid in taming the reactions of molecular
oxygen (and its peroxide relatives) with organic reductants. The
hope is to develop new processes for the production of desirable
oxidation products.

In addition to molecular oxygen itself, its one (superoxide)
and two (peroxides) electron reduction products are likely candidates
for being useful oxidants. Considerable utility has already been
demonstrated for alkyl hydroperoxides and hydrogen peroxide. Like
molecular oxygen, the peroxides offer the advantage of leaving be-
hind innocuous by-products. This is a distinct advantage for any
oxidant, since there is mounting concern over the disposal of toxic
wastes.

The discussion to follow deals with four areas:
A. Direct Activation of Molecular Oxygen for Nonradical Oxida-
tions.
B. Selective Radical Oxidation Processes.
C. Selective Oxidations with Peroxides.

D.   Indirect Oxidations

Metal catalysts figure heavily in all four of these areas.
Metallic centers offer unique possibilities for increasing selectivity in redox processes. One of the most common roles played by metal
catalysts is bringing the reactants together in the same coordination
sphere; and, in the words of R. B. Woodward, "Enforced propinquity
often leads on to greater intimacy."

Although the first priority for research in homogeneous catalytic oxidations should be the discovery of new reactivity, another
likely and important ramification seems worthy of mention at this
time. Since metallic catalysts almost always have coordination
sites available which are not directly involved in the catalytic process, one should be able to attach ligands with the appropriate
molecular architecture for creating a catalytic center having enzyme-
like specificity for substrate recognition. Catalysts constructed
in this way should offer unique opportunities for selective (e.g.,
enantioselective or regioselective) oxidations. In contrast to the
impressive accomplishments in catalytic asymmetric reductions, catalytic asymmetric oxidations are unknown.

Finally, it is hoped that the study of homogeneous catalytic
oxidations will improve understanding of the analogous heterogeneous
catalytic processes.

A.   Direct Activation of Molecular Oxygen for Nonradical Oxidations

The interaction of molecular oxygen with reduced transition
metals affords two principal types of mononuclear adducts:

Type I (side-on)   Type II (end-on)

e.q. $Ni°$, $Pd°$, $Pt°$, Ir^I, Rh^I e.q. Fe^{II}, Co^{II}

Several examples where the adducts of type I are involved suggest
that oxidations may occur with the substrate and the oxygen coordinated to the same metal:

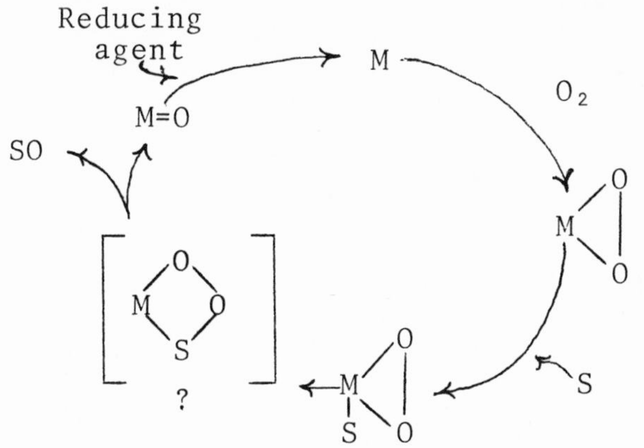

Reducing
agent ⟶ M
SO
M=O
O$_2$

[M, O, S] ?

A mechanism with a similar coordination-insertion sequence has been proposed for the epoxidation of olefins by a peroxide complex of molybdenum, and for the insertion of CO, CO_2, aldehydes, ketones,

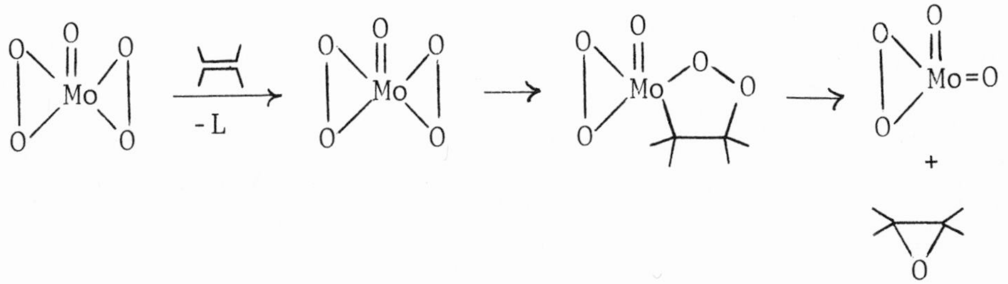

and electrophilic olefins into the Pd and Pt oxygen complexes:

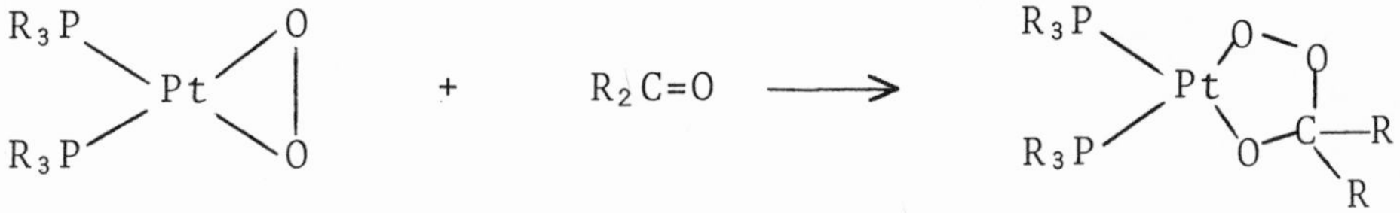

$$R_3P,Pt,O + R_2C=O \longrightarrow R_3P,Pt,O-C-R,R$$

A similar mechanistic sequence can also be proposed for the recently reported, rhodium catalyzed, selective oxidation of terminal olefins to methyl ketones. To make such a reaction catalytic it is necessary to use a coreducing agent, for example a phosphine, to regenerate the reduced state of the metal.

With the possible (if it involves a Rh(olefin)O_2 complex) exception of the rhodium catalyzed oxidation of olefin to methyl ketone,

these type I oxygen complexes of the noble metals have proved re-
markably inert toward <u>interesting</u> organic substrates such as olefins.
We mention this here because many of the workers in this area had to
learn this fact the hard way, since the negative results of others
were not published. It is hoped that chemists will no longer waste
their energy in trying to make L_2PtO_2 react with olefins.

The complexes of type II are mainly involved in the biological
monooxygenases and their models. The oxidations achieved by these
systems are principally of the mixed function type:

$$S \; + \; DH_2 \; + \; {}^*O_2 \; \longrightarrow \; SO^* \; + \; D \; + H_2O^*$$

In the nonenzymic models for these oxygenases a variety of reducing
agents have been employed. In one of the more successful recent
model systems, hydroxylation of cyclohexane was accomplished using
Fe^{III}, O_2, and $\phi NHNH\phi$. In serving as the reducing agent the hydrazo-
benzene was oxidized to azobenzene. In spite of some progress toward understanding the chemistry of monooxygenases much work remains
to be done. Of course, the eventual goal is to apply this chemistry
to practical oxidation problems. At present, this goal is far from
being realized.

A common reaction catalyzed by dioxygenases is the cleavage of
carbon-carbon double bonds. Reactions which at least appear to be
models for such transformations have been observed for activated
olefinic linkages:

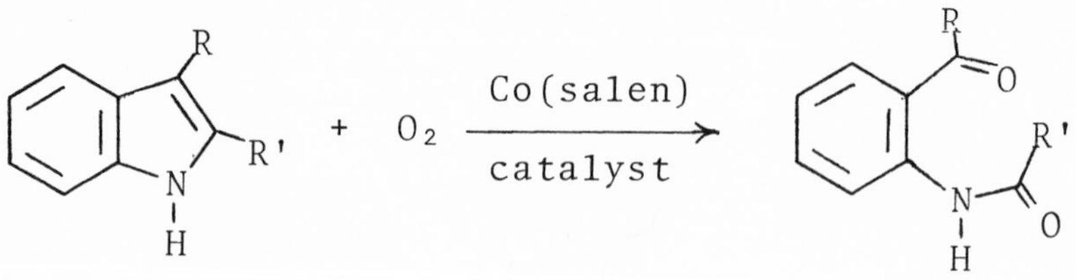

The examples presented here reveal that oxygen activation can
occur in several different ways. However, the actual mechanisms
of these reactions and the nature of activated O_2 remain obscure.
It is thus important to continue the effort to clarify the properties
and the reactivity of oxygen adducts. At the same time that this
more fundamental research is going on, it will also be important to
continue prospecting for catalytic systems which are suitable for
practical (i.e., large scale) production of oxychemicals.

When thinking of how to activate O_2 one thinks almost exclusively of transition metals. There is at least one indication that such
an approach may be too restrictive. It has recently been shown that

a singlet oxygen type reaction can be realized with ground state oxygen in the presence of catalytic amounts of certain radical cat- ions:

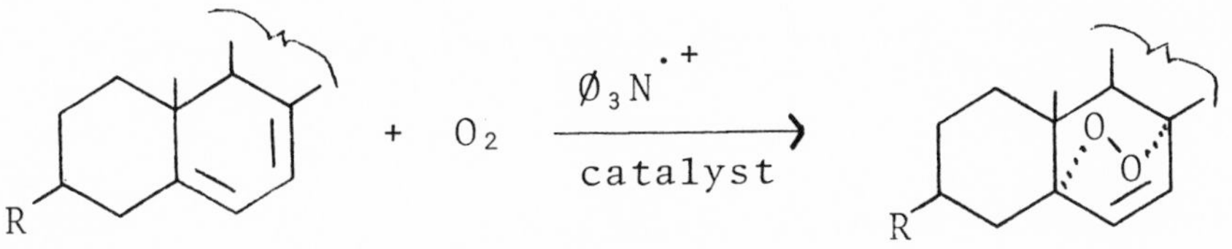

B.   The Influence of Metal Complexes on the Course and Selectivity
                of Radical Initiated Oxidation

     The attempt to catalytically activate molecular oxygen toward direct interaction with a hydrocarbon is often unsuccessful and the organic compound oxidizes by a free radical initiated autoxidation pathway instead. In many cases autoxidation in the absence of metal

$$R-H \xrightarrow[-InH]{-In\cdot} R\cdot \xrightarrow{O_2} RO_2^{\cdot} \xrightarrow[-R\cdot]{+RH} ROOH \longrightarrow \begin{array}{l} \text{stable} \\ \text{oxidation} \\ \text{products} \end{array}$$

complexes tends to be unselective and not particularly useful.

     Free radical initiated autoxidations carried out in the presence of certain transition metal complexes, however, are highly selective and show promise of becoming synthetically useful. Furthermore, the products of free radical autoxidations are highly dependent on the nature of the metal complex present in solution. For example, the product profile of radical initiated olefin autoxidation is markedly dependent upon the metal complex used as the catalyst. In pentane autoxidation, a unique selectivity has been observed for the formation

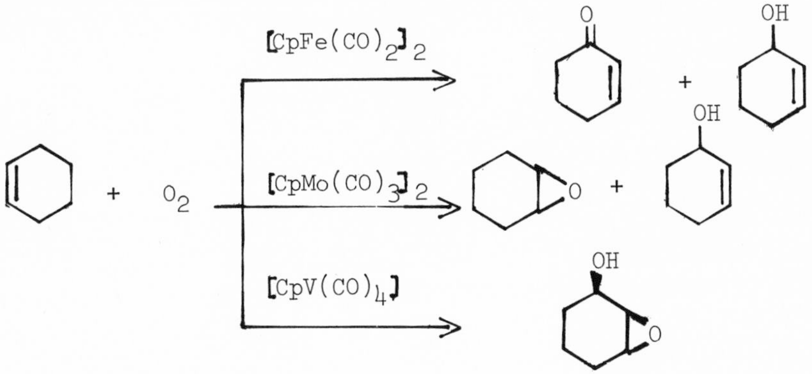

2-pentanone in good yield (80%). It appears that the function of the metal complex in these cases is to intercept a reactive intermediate formed during autoxidation and to convert it in a selective

$$O_2 \quad + \quad CH_3CH_2CH_2CH_2CH_3 \xrightarrow[40^\circ C]{Co(III)} \quad CH_3\underset{O}{\overset{\parallel}{C}}-CH_2CH_2CH_3$$

manner to stable oxidation products.

The moderating influence of the metal complex on autoxidation selectivity suggests that a greater degree of control of autoxidation might be obtained if we better appreciated the specific nature of the interactions of metal complexes with the intermediates of autoxidation (R, ROO, ROOH, $(C-OO)_x$, etc.). Several areas (a-e) in which metals might direct the course of autoxidation, could benefit from systematic investigation.

1. Metal Promoted Generation of Alkyl Radicals in situ and their Selective Interaction with Dioxygen. Hydrocarbons react with a wide variety of metal complexes to give metal alkyls by a number of different pathways:

oxidative addition $RH + M^{n+} \longrightarrow R-M^{n+2}-H$
electrophilic substitution $RH + M^{n+1} \longrightarrow R-M^{n+} + H^+$
addition to double bonds $C=C + MH \longrightarrow RM$, etc.

Complexes in which metal-alkyl bond strengths are sufficiently weak may exhibit radical-like reactivity. It might be expected that radicals generated from the homolytic cleavage of some metal alkyls would exhibit interesting selectivity since departure of the radical from the coordination sphere may not be complete prior to oxidation.

$$M-R \longrightarrow M\cdot \quad + \quad R\cdot \xrightarrow{O_2} \quad M\cdot \quad + \quad RO_2\dot{}$$

Metal alkyls with stronger carbon-metal bonds may interact directly with dioxygen via an insertion reaction to form metal peroxy complexes. It is known, for example, that alkyl cobaloxime complexes are capable of inserting dioxygen between the metal center and the

$$M-R \quad + \quad O_2 \longrightarrow MOOR$$

alkyl group. Little is known, however, concerning the reaction of dioxygen with other metal alkyls or the stability and reactivity of metal hydroperoxy complexes.

2. Metal-Moderated Interactions of Alkyl Radicals with Dioxygen. Reactions of paramagnetic complexes with radicals are well known. For example, benzyl radicals react with Cr(II) to give benzyl chrom-

ium(III) in high yield. Significant efforts have not been made to
use such reactions as a means of moderating the reactivity of free
radicals with dioxygen. It should be noted that formation of metal

$$R\cdot \;+\; M^n \longrightarrow M^{n+1}\text{-}R \xrightarrow{\;O_2\;}$$

alkyls by capture of highly reactive radical species which may be
present in low concentration presents a statistical problem which
necessitates relatively large metal concentrations.

3. Reactions of Metal Complexes with Hydroperoxides, Peroxy
Radicals, and Polyperoxides. Metal hydroperoxy complexes can be
generated not only from metal alkyls as discussed above but also
from interaction of other autoxidation intermediates with the metal
center. Problems of both a statistical as well as a chemical nature
could mitigate against capture of ROO· by a metal center, however,
reactions of ROOH with the metal should be studied in detail. A
subsequent section will deal specifically with selective oxidations
using ROOH in the presence of group V and VI metal centers. A number
of group VIII metal complexes react with hydroperoxides to form alkylperoxy complexes; however, little is known concerning their reactivity or stability. Reactions of metal alkylperoxy complexes with acids

to give hydroperoxides or reaction with unsaturated organic substrates
to give organic oxidation products is an area which might prove to
be most fruitful. Such studies may suggest modification of autoxidation reactions in a beneficial manner.

Since hydroperoxides are the intermediates of many radical in-
intiated autoxidation reactions, any approach which would promote
the rapid generation of hydroperoxides but which would stabilize them
for prolonged periods during which they might undergo selective reaction to more stable oxidation products, would be highly desirable.

The rapid catalytic decomposition of alkyl polyperoxides formed
during autoxidation of some olefins in the presence of complexes
such as $RuCl_2(PPh_3)_3$ gives rise to much higher yields of double
bond cleavage products than are normally obtained. Alpha olefins
give carboxylic acids or aldehydes in good yields with epoxides as by-
products. There is evidence suggesting that polyperoxides decompose
directly to epoxides in some instances. Almost no data are available

on the catalytic decomposition of polyperoxides by metal complexes in solution.

4. Metal Catalyzed Destruction of Unwanted Side Products by Selective Total Oxidation. Under appropriate conditions, the autoxidation of propylene to propylene oxide in the liquid phase can be carried out with selectivities approaching 60%. One of the major problems with the utility of this method is the separation of propylene oxide from the large number of by-products. Another problem is the reactivity of propylene oxide with acidic and hydroxylic by-products. Carboxylic acids and alcohols selectively complex with many metal centers including molybdenum or vanadium catalysts in preference to epoxides and thus might be selectively (and sacrificially) destroyed by complete catalytic oxidation to smaller molecules and eventually CO_2. Research in this area could conceivably produce a cleaner product stream far richer in propylene oxide than in the absence of catalyst.

5. Studies of Metal-Promoted Oxidations Under Conditions of Severe Radical Inhibition. In addition to attempting to cope with radical reactions by increasing their selectivity, it might be desirable to investigate catalytic reactions of organic compounds known to proceed predominantly by radical pathways under conditions of severe inhibition. Such an approach may uncover interesting leads regarding non-radical pathways which have been heretofore ignored due to predominance of radical pathways. If identified, these pathways might be optimized. Some success along these lines has apparently been realized with olefin oxidations in the presence of rhodium(i), however, other autoxidations have not been treated in this manner. As shown in the first section of this report, Rh(I) can catalyze selective oxidation of 1-olefins to methyl ketones under conditions of severe radical inhibition.

C. Selective Oxidations with Peroxides

Catalytic reactions of peroxides involving both homolytic and heterolytic mechanisms are known. The first type involves transition metal catalysts such as Co, Mn, Fe and Cu which react with peroxides via one-electron transfer reactions producing intermediate radicals, e.g., with an alkyl hydroperoxide:

$$RO_2H + M^{n+} \longrightarrow RO_2\cdot + M^{(n-1)+} + H^+$$

$$RO_2H + M^{(n-1)+} \longrightarrow RO\cdot + M^{n+} + OH^-$$

These reactions are generally characterized by low selectivity and are limited in their applicability (see Section 2 of this report). The second group involves oxygen transfer in a two electron reaction and these reactions are, in general, much more selective:

$$S \ + \ RO_2H \xrightarrow{\text{catalyst}} SO \ + \ ROH$$

In these reactions coordination of the peroxide to electrophilic reagents, such as Bronsted and Lewis acids including transition metals in high oxidation states (Mo^{VI}, W^{VI}, V^V and Ti^{IV}), renders the peroxidic oxygen more electrophilic and, hence, more susceptible to reactions with nucleophilic substances, such as olefins.

The two most important classes of peroxides from an economic point of view are hydrogen peroxide and alkyl hydroperoxides. Hydrogen peroxide has the important advantage that the co-product of oxidation is water, thus presenting no recycle or disposal problems. Moreover, it has been shown that hydrogen peroxide can also be pro-

$$S \ + \ H_2O \xrightarrow{\text{catalyst}} SO \ + \ H_2O$$

duced directly from hydrogen and oxygen in the presence of a transition metal catalyst:

$$H_2 \ + \ O_2 \xrightarrow{\text{catalyst}} H_2O_2$$

The major problem associated with the utilization of hydrogen peroxide is the fact that it is produced as an aqueous solution (generally 30% or less) and reactions have to be carried out in polar, coordinating solvents. Under these conditions, reactions are generally slow and electrophilic catalysis is inefficient due to competing coordination of the polar solvent. Moreover, facile ligand exchange in these media imposes severe limitations on the type of ligands that can be used.

One approach to circumventing this problem could utilize phase transfer catalysis in order to bring the hydrogen peroxide (in some form) into a non-polar solvent. Because of the limitations discussed above, hydrogen peroxide has actually limited application as an oxidant, and is generally used only for oxidation of strongly nucleophilic substrates, such as sulphides and amines.

The major objectives in this area remain the mild selective epoxidation of simple olefins and hydroxylation of C-H bonds with hydrogen peroxide.

Alkyl hydroperoxides, such as tert-butyl and ethylbenzene hydroperoxides, are commercially attractive oxidants that are readily prepared by autoxidation of the corresponding hydrocarbons (or from hydrogen peroxide). They exhibit similar properties to hydrogen peroxide and possess the added advantage of solubility in non-polar media. Electrophilic catalysis is, therefore, much more efficient and epoxidation of simple olefins is feasible as in the commercial process for the epoxidation of propylene in the presence of a high valent transition metal catalyst (Mo^{VI}, W^{VI}, V^{V}, Ti^{IV}):

$$RO_2H \quad + \quad \text{olefin} \quad \xrightarrow{\text{catalyst}} \quad ROH \quad + \quad \text{epoxide}$$

A major drawback to the commercial use of alkyl hydroperoxides as oxidants is, however, the unavoidable formation of the alcohol co-product, which must be either recycled or disposed of (sold). Hence, the continual incentive for the development of analogous processes utilizing hydrogen peroxide.

Up until now, these metal catalyst-alkyl hydroperoxide systems have been used only for the epoxidation of simple olefins and amines and sulphides. A desirable objective is the extension of the use of these reagents to the synthesis of more complex molecules. The transition metal catalyst, by virtue of its complex coordination sphere, should impose steric constraints which profoundly influence the stereochemistry of these reactions thus permitting both regio- and stereoselective control. A few examples are known which demonstrate that both regio- and stereoselective reactions are feasible with these reagents. For example, geraniol is regioselectively epoxidized to the previously unknown mono-epoxide with tert-BuO_2H/ $VO(acac)_2$ catalyst:

$$\text{geraniol} \quad \xrightarrow[\text{tert-BuO}_2\text{H}]{\text{VO(acac)}_2} \quad \text{mono-epoxide}$$

An example of a stereoselective reaction is epoxidation of the homoallylic alcohol below which afforded only the syn-epoxy alcohol with tert-BuO_2H/$Mo(CO)_6$ catalyst:

$$\text{homoallylic alcohol} \quad \xrightarrow[\text{tert-BuO}_2\text{H}]{\text{Mo(CO)}_6} \quad \text{syn-epoxy alcohol}$$

The application of these reagents to the regio- and stereoselective synthesis of a variety of complex molecules (fine chemicals, pharmaceuticals, etc.) is worthy of further attention.

D. Indirect Oxidations

There is a general class of selective oxidations, many involving dioxygen, which may be termed "indirect" in that a metal complex oxidizes a substrate and then the reduced complex (catalyst) is re-oxidized by O_2. A well known example is oxidation of ethylene to acetaldehyde by chloropalladate(II) in aqueous acid media (Wacker Process); a two-electron redox process with a net incorporation of an oxygen atom from the medium into the ethylene leaves the palladium in a zero oxidation state, and this is reoxidized using an $O_2/Cu(II)$ system. Ascorbate oxidase uses enzyme-bound copper(II) to oxidize ascorbic acid to dehydroascorbic acid, and the resulting copper(I) is subsequently reoxidized using O_2.

Other metal ion two-electron oxidants (e.g., Rh(III), Tl(III), Hg(II)) have been studied. The rhodium(III$\rightleftharpoons$I) cycle can be maintained using O_2 and the interesting coupling of such a Wacker cycle to an "O_2 activation cycle" could lead to utilization of both oxygen atoms of the O_2 molecule for ketone production. The thallium(III) Wacker type system gives, in addition to carbonyl products, glycols; but the in situ reoxidation of thallium(I) is not accomplished with O_2. The production of organic products via the "Oxymercuration" reaction again can give stoichiometric olefin oxidation and O_2 - resistant mercury (0). Certain advantages exist in separating, the oxidation and catalyst reoxidation processes, e.g., in the application of supported catalyst (column) procedures. There is a wide range of stoichiometric "oxypalladation" and coupling reactions involving organics and palladium(II); these are formally oxidation reactions and the oxidant is reduced to Pd (0). However, the conditions necessary for metal reoxidation are fairly restrictive and commonly the acid conditions of the Wacker Process cannot be tolerated. Other reoxidation systems, not necessarily in situ ones, should be investigated, and these could be chemical (e.g., H_2O_2, Cl_2, quinones, nitric acid, NO_2, SO_2) or electrochemical.

The conversion of benzene to phenol via acetoxypalladation could be usefully made catalytic.

Requirements of a metal ion for effective olefin oxidation are essentially two-fold: the ion must be able to coordinate the olefin and then undergo a two-electron reduction. However, these are non-complimentary properties in that low valence metal species favor olefin coordination, while the oxidizing capabilities lie with high valency species. Alternative pathways involving chemical oxidation of a lower valence olefin complex (e.g., Ru(II), Pt(II)) to a reactive higher valence intermediate should be considered.

The selective oxygen atom incorporation <u>via</u> H_2O is not restricted to olefinic substrates and other organic and inorganic unsaturates are potential candidates. An example of the latter is carbon monoxide, and the system is formally similar to the ethylene one: reductive carbonylation of the metal center (e.g., Co(III) Rh(III)) leads to CO_2 and univalent metal species. Reoxidation of these using O_2 leads to catalytic conversion: $2CO + O_2 \rightarrow 2CO_2$, but <u>any</u> oxidant capable of reoxidizing the low valent metal species may be catalytically reduced using CO (or C_2H_4). In any substrate system, the selective oxygen atom transfer effectively provides a two electron source, and the chemistry of the indirect oxidations centers around the solvent water molecule:

$$H_2O \longrightarrow O + 2H^+ + 2e^-$$

The $(2H + 2e)$ system instead of reducing the metal center could alternatively be used as a dihydrogen source, e.g.,

$$\text{substrate} + H_2O \longrightarrow \text{Oxidized substrate} + H_2$$

The net reaction for an olefinic substrate being hydration followed by dehydrogenation. (The case of CO as substrate gives the familiar water-gas shift reaction). Methylethyl ketone production from 1-butene <u>via</u> such a process has been demonstrated, and is effected using a heterogeneous catalyst, although forcing conditions are required and side-products result. Development of effective and selective catalysts for such processes seems highly desirable.

In this area of indirect oxidation, as in other areas of oxidation, there has perhaps been too much emphasis on transition metals, it would seem worthwhile to consider investigating main group elements as catalysts. The recently developed tellurium catalyzed conversion of ethylene to ethylene glycol is an example of such a process.

E.   Conclusions and Recommendations

As the hydrocarbon raw materials become increasingly dear, the need for more selective processes for transforming them to useful oxychemicals becomes more apparent. Molecular oxygen and to a lesser extent hydrogen peroxide are very attractive oxidants both from the standpoint of cost and because of the innocuous nature of water, their reduction product. Thus great emphasis should be placed on research which aims at coupling important oxidative transformations to these two oxidants. The following reactions are examples of desirable oxidative transformations which have not yet been accomplished using either O_2 or H_2O_2 as the oxidant:

1. $CH_2=CHCH_3 \longrightarrow$ (propylene oxide)

2. $CH_2=CH-CH=CH_2 \longrightarrow$ HO–CH$_2$–CH=CH–CH$_2$–OH

3. (benzene) $\longrightarrow$ (phenol) OH

4. (cyclohexene) $\longrightarrow$ CHO / CHO (open chain dialdehyde)

5. $R-CH_2-CH_2-CH_3 \xrightarrow{\ \ } R-CH_2-CH_2-COOH$

6. $CH_2=CH-CH_2-CH_3 \longrightarrow$ CH$_3$-CO-CH$_2$-CH$_3$ (methyl ethyl ketone)

Of course some transformations (e.g., 3 and 5) have never been achieved (with high selectivity) using any oxidant. Thus, although O_2 and H_2O_2 may be the ideal oxidants, it should be apparent that many useful processes do now and will in the future function profitably using less convenient oxidants.

Metal catalysts figure prominently in most of the approaches to developing selective oxidation processes. Hence it will probably be important to devise ligands which bind strongly to the metal center, and which resist attack by the oxidizing milieu in which they must necessarily function.

III. CARBON MONOXIDE REACTIONS

Co-Chairmen: D. Forster, F. Piacenti
Members: U. Belluco, G. Braca, F. Calderazzo, A. Ligorati,
I. E. Ruyter, J. Stille, H. Tkatchenko, J. Tsuji,
M. Tsutsui

The ready availability of carbon monoxide on a large scale together with its reactivity towards transition metals and its facile
insertion into metal-carbon bonds has made it an attractive material
for industrial organic synthesis. In fact, reactions involving carbon monoxide represent the largest use of homogeneous catalysis.
The future of carbon monoxide as a key material in organic chemical
synthesis seems to be assured since it can be produced on a commercial scale from many different sources of carbon. Thus, it has been
manufactured by both steam reforming and partial oxidation of methane,
naphtha, coal and also by reduction of carbon dioxide. It has the
potential of being the building block between the fundamental forms
of carbon and more sophisticated organic molecules via Fischer-Tropsch
and other building reactions.

In this report we examine the state-of-knowledge and the future
potential for the principal reactions involving carbon monoxide.

A. CO Production, Purification and Concentration

The production of CO by combustion and steam reforming techniques appears to be only amenable in the foreseeable future to catalysis at high temperature using heterogeneous catalysts. We, therefore,
do not recommend that much work be pursued in this area.

However, there are a number of sources of CO which are presently
not used for chemical production, e.g., blast furnace off-gases,
phosphorus furnace gases, and the ferro-silicon industry. The problem with these sources is that they contain substancial proportions of
diluents and impurities. A concentration and purification technique
for their streams may represent a significant advance. Therefore,

an improved technique for recovery of CO by adsorption-desorption
cycle would be valuable. The present processes for absorption using
copper-complexes suffer from a number of disadvantages, particularly
water susceptibility and loss of ligand.

The generation of CO from CO_2 may be an area which is amenable
to a photocatalytic approach.

B. Water-gas Shift Reaction

The present heterogeneous catalytic systems for performing the
water-gas shift reaction have the disadvantage of operating at high
temperatures where the equilibrium is not favorable. A homogeneous
catalyst which could operate efficiently at low temperatures and low
CO pressures would represent a substantial advance because this could
alleviate the problem of operating a multi-stage reaction with CO_2
absorption to force the equilibrium.

C. Reduction of Carbon Monoxide

The reduction of carbon monoxide is important in methane, methanol, and Fischer-Tropsch synthesis and potentially in formaldehyde
synthesis. The methane, methanol, and Fischer-Tropsch synthesis
reactions are frequently only achievable efficiently with heterogeneous catalysts which, however, are subject to a number of disadvantages (e.g., sulfur poisoning and carbide formation). There
could be considerable advantages to be gained by conducting these
reactions in the liquid phase because of the very exothermic nature
of the processes. There is a decided lack of knowledge concerning the
fundamental reaction of CO with H_2. We recommend research which can
identify the mechanistic pathways for both the heterogeneously and
homogeneously catalyzed reactions of CO with H_2. The recent report
of CO/H_2 reduction to CH_4 in the presence of cluster carbonyls in
solution may be a starting point for studying these fundamental
processes. The knowledge gained should be valuable in providing an
understanding of the heterogeneous processes also.

In the area of methane, methanol, and formaldehyde synthesis, we
recommend research which can identify and the mechanistic pathways
of CO/H_2 reactions both on heterogeneous and homogeneous catalysts.
We recommend research on developing homogeneous catalysts for all of
these reactions. Chemical and spectroscopic studies are desirable
of the effects of ligands and promoters on the metal-carbon monoxide
bonding, which may in turn affect hydrogen mobility in hydrido-carbon-
yls.

With respect to the Fischer-Tropsch synthesis, much of the effort has been directed towards the production of C_5-C_{11} straight

chain hydrocarbons for fuel use. Recent developments in the area
of generating fuels from short chain materials such as methanol and
olefins using olefin dimerization catalysts, particularly with homogeneous Ni/Al catalysts which give highly branched olefins, may require a modification of our thinking with respect to the required
products of the Fischer-Tropsch synthesis. In particular, it makes
it feasible to operate a Fischer-Tropsch synthesis under conditions
where short chain olefins are the main products and then combine
this reaction with a dimerization catalyst either to generate a high
octane fuel or as a feedstock for chemical synthesis.

D. Ethylene Glycol Synthesis

The rhodium cluster carbonyl species which has been described
by Union Carbide for glycol synthesis from CO/H_2 is probably the
most outstanding example of the use of a metal cluster in homogeneous
catalysis. Fundamental knowledge gained here could have widespread
impact on using homogeneous catalysts in the selective synthesis of
oxygenated compounds of higher molecular weight from carbon monoxide.
Spectroscopic studies of the metal species either under actual reaction conditions or conditions relevant to the reaction should give
valuable information on the mechanistic pathways and may help in
devising systems with lower pressure requirements. The chemistry
of metal-metal bonded species should be investigated, particularly
with regard to carbon-carbon coupling (see section V-C on Metal
Clusters in Catalysis).

E. Hydroformylation

The working conditions and selectivity have been improved greatly
by the introduction of rhodium-phosphine based catalysts. However,
the reaction rate is still relatively slow as compared to rhodium-
catalyzed hydrogenation reactions. If the reactivity could be substantially improved, it would allow a further reduction in temperature which would in turn allow improvements in asymmetric synthesis
using these systems. A definitive mechanistic study of the rhodium
hydroformylation reaction, incorporating in situ spectroscopic techniques, is desirable.

A disadvantage of the rhodium system is that it is not an isomerization catalyst and, hence, gives mainly branched aldehyde from
internal olefins. A dual functional system which combines a very
discriminating and reactive rhodium hydroformylation catalyst with
an isomerization catalyst would be a significant development (See
section VI). The isomerization catalyst would have to be capable
of working very efficiently in the presence of excess ligand and
carbon monoxide.

Cobalt catalysts appear to be capable of further improvement, particularly with regard to improving selectivity and rate.

The development of milder and stereospecific hydroformylation reaction conditions has created the possibility of reactions of olefins bearing other functional groups which has increased the utility of this area in poly-functional organic compounds, e.g., the synthesis of glutamic acid by a route involving hydroformylation of acrylonitrile. Further investigations to increase the scope of this area should be fruitful.

F. Hydrocarboxylation and Related Reactions

The only transition metal catalyzed, large scale hydrocarbonylation process involves synthesis of propionic acid from ethylene using a nickel carbonyl catalyst. The reaction conditions are very severe. Palladium and cobalt catalyst systems are known to operate under much milder conditions. Fundamental information about the intimate mechanistic detail is lacking. The palladium-phosphine catalyst system is capable of asymmetric synthesis, and improvement in the reactivity of this system could give substantial improvement in the optical yield obtained. Palladium has also shown the ability to produce dicarboxylic acids from mono-olefins. The problem with this reaction is regeneration of Pd(0), since the approaches taken to date generate water which inhibits the catalyst. Fundamental approaches to Pd(0) oxidation are recommended to attempt to circumvent this problem. Since both the hydrocarboxylation and dicarboxylation reactions are also stereospecific, low temperature processes, a study of the application of these reactions to the synthesis of multifunctional compounds is also warranted.

G. Alcohol Carbonylation and Homologation

Two large scale processes are being operated for carbonylation of methanol to acetic acid; one is based on a homogeneous cobalt catalyst and the other is based on rhodium. Both systems require an iodide promoter. The rhodium system operates under relatively mild conditions. A system which did not require a halide promoter would be a marked improvement.

The homologation of methanol to ethanol has been observed in the presence of cobalt carbonyl. This reaction is a potentially important method of generating C_2 molecules. Ethylene is the principal hydrocarbon used by the chemical industry and is produced currently in large quantities from natural gas. In the homologation reaction with the cobalt catalyst the selectivity is not very good. Further, the mechanism of this reaction is not well understood, although it appears that the high acidity of $HCo(CO)_4$ is of key importance.

This acid function of $HCo(CO)_4$ is not well understood and deserves further study. Other metal systems and ligands should be investigated for activity in the reaction. If the acidity of the $HCo(CO)_4$ proves difficult to duplicate with other metal-hydrido species then a dual functional system may be warranted incorporating for example a rhodium catalyst and an acid catalyst.

H.   Acetylene Carbonylation

Several large commercial plants have been operated for the production of acrylic acid derivatives from acetylene using nickel carbonyl catalysts and a halide promoter. The actual metal species involved in the catalytic cycle have not been determined but some visual observation has suggested that species other than simple mononuclear complexes are involved. The role of the halide promoter has not been completely elucidated. A more detailed investigation of this catalyst is desirable.

Cyclization by combination of acetylene and carbon monoxide has been recognized but although an interesting synthetic area it suffers from the problems of low selectivity, low productivity and low activity. Wider knowledge and improvement of the catalysts is desirable.

I.   Carbonylation of Aromatic Compounds

The catalytic carbonylation of aromatic substances other than halides with metal complexes has not been accomplished. A stoichiometric technique has been developed in which a combined mercury-palladium system is used; however, palladium(O) is precipitated. Several fundamental problems have been identified with the use of palladium in aromatic carbonylation, e.g., slow reaction with aromatic hydrocarbons and formation of palladium(O) in the catalytic cycle. Fundamental studies aimed at providing more efficient aromatic activation systems are desirable. Direct complexation of the aromatic system on cluster carbonyl species might provide an opportunity.

J.   Carbonlyation of Saturated Substrates

Ethers and epoxides are carbonylated in presence of $HCo(CO)_4$ and the mechanism of the reactions are quite well understood, with the acid function of $HCo(CO)_4$ serving to facilitate the formation of the metal-carbon bond. The selectivity of the reaction is rather poor in some cases and the use of the appropriate ligands might improve the selectivity. Alternatively other metal-hydrido species should be investigated both in the catalytic reactions and also in stoichiometric reactions with species such as epoxides and other ethers.

K.   Carbon Dioxide

There have been several reports in which CO_2 has been introduced into an organic molecule <u>via</u> use of a transition metal complex It has been reported that ruthenium hydride can insert CO_2 to give a formate complex and a rhodium aryl complex reacted with CO_2 to give a benzoate complex. Other reports of metal-CO_2 complexes have appeared. Further fundamental studies of the condition and chemical environments necessary for CO_2 fixation via transition metal systems are desirable.

L.   Carbon Monoxide as Reducing Agent

Reduction of nitrobenzene to aniline by carbon monoxide has been accomplished using ruthenium catalysts in the presence of hydrogen. Reduction of nitro compounds to isocyanate could be a valuable area to be investigated.

M.   Miscellaneous Reactions

Several reports have appeared where carbon monoxide insertion into a metal-nitrogen bond appears to have occurred. The mechanistic pathway of these reactions, however, does not appear clear at all. Investigations in this field appear profitable.

Little is known about the insertion of carbon monoxide into the oxygen-metal bond. A deeper insight in the course of this reaction may be useful in view of a better understanding of the synthesis of formates and of the water-gas shift reaction. The electrolytic polymerization of carbon monoxide to squaric acid involves some mechanistic steps which might be relevant to obtaining an understanding of the Fischer-Tropsch synthesis.

Cyclic oxygenated compounds may be synthesized from olefins or diolefins with carbon monoxide to produce ketones. Further work to increase the general utility of this type of reaction appears desirable.

Conclusions and Recommendations

1. Many reactions involving carbon monoxide are highly exothermic and the problems of heat removal in such reactions could be considerably alleviated by operating in the liquid phase, and therefore

catalysts capable of operating at lower temperature in the liquid phase are highly desirable. Homogeneous catalysts should provide such an opportunity.

2. Considerable progress has been made in the understanding at the fundamental level of carbon monoxide reactions on transition metals. However much remains to be done and should prove very valuable as an aid to designing catalytic systems with improved reactivity and selectivity.

3. Fundamental studies, particularly in situ spectroscopic studies, of the reactions of CO and hydrogen over metal carbonyl complexes appear highly desirable.

4. Continued studies are warranted of the influence of ligand structure and solvent properties on the reactivity and selectivity of metal complexes in carbonylation reactions.

5. Research directed toward the use of carbon monoxide in stereospecific and asymmetric synthesis of organic compounds, especially when such compounds contain functional groups in addition to the introduced carbonyl moity, is recommended.

Bibliography

1. "Carbon Monoxide in Organic Synthesis," J. Falbe; Springer-Verlag, 1970.
2. "Organic Synthesis with Metal Carbonyls," Vol. II, Ed. Wender and Pino, Interscience 1977.
3. "Transition Metals in Homogeneous Catalysis," Ed. G. N. Schrauzer, Marcel Dekker, 1971.
4. "Aspects of Homogeneous Catalysis," Vol. 2, Ed. R. Ugo D. Reidel 1974.
5. I. Wender, Catalysis Reviews, 14, 97 (1976).
6. F. E. Paulik, Catalysis Reviews, 6, 49 (1972).
7. M. Orchin and W. Rupilius, Catalysis Reviews, 6, 85 (1972).

IV. FRONTIER AREAS BETWEEN HOMOGENEOUS AND HETEROGENEOUS CATALYSTS

Co-Chairmen: J. M. Basset, J. Norton
Members: G. Cainelli, P. Chini, G. Dolcetti, N. Giordano,
          M. Graziani, R. H. Grubbs, R. Leoni, M. Rossi,
          G. Sbrana, R. Ugo

A number of suggestions have been made for the extension of the concepts of homogeneous catalysis into new areas. Its very nature creates difficulties: reactants, products, and catalysts are all in one phase. Hence the use of insoluble supports has been suggested, as has the use of two liquid phases along with phase transfer reagents. Furthermore, interest in homogeneous catalysts containing more than one metal has arisen partly because of an apparent analogy between these compounds and heterogeneous metal surfaces.

A. CATALYSIS BY SUPPORTED COMPLEXES

The attachment of a homogeneous catalyst to a solid support provides a technique for the easy recovery of a homogeneous catalyst. Two main approaches have been studied so far (see Figure).

1. Grafting the catalyst onto a support with the aim of keeping its molecular nature. Most studies using this approach have involved polymers and oxides.

2. Grafting the catalyst onto a support without trying to preserve its molecular nature. Most of these studies have been carried out on oxides.

It has been amply demonstrated that the first approach permits any homogeneous catalyst to maintain its basic catalytic activity and selectivity, although the support may superimpose additional effects. Emphasis in the future should be on the control and exploitation of these effects.

215

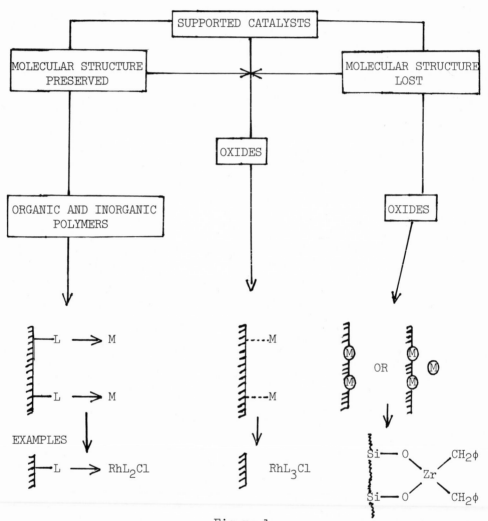

Figure 1.

The second approach produces unexpectedly high increases in activity and/or selectivity compared with homogeneous precursors. There has thus been produced a new class of heterogeneous catalysts of high industrial importance. The following sections identify areas for study within each of these two approaches.

1. The Molecular Nature of the Complex is Preserved

a. Supports. It appears now that the choice of the support is closely related to the catalytic reaction which has to be carried out. New polymers (with a known polarity) as well as chemical modifications of an oxide surface should be examined to determine the correct environments for supported complexes. Thermal and chemical stability of supports in catalytic media is also important. Finally, mechanical and thermal properties (in case of exothermic reactions) and proximity of binding sites seem to be key factors in the choice of supports.

Important new types of supports which merit attention include zeolites, clays, and glasses. Their regular topological structure could be used for the entrapment of the catalyst and thus for the modification of its selectivity.

Electron-conducting supports such as graphite and semiconductors are of potential interest for catalyzed redox reactions. The mode of attachment should maintain facile electron flow.

b. Kinetic Problems. With a specific choice of support polarity, solvent, and substrate supported complexes may exhibit activity and selectivity which are comparable to those obtained in pure homogeneous catalysis (e.g., asymmetric synthesis).

The following topics need fundamental research:

1. Very precise kinetic measurements are needed to compare the activity of a given supported complex with that of its unsupported equivalent. Care should be taken to ensure the similarity of the ligand groups, and a catalytic reaction as simple as possible should be studied. Such research should also cover the influence of systematic loading on catalytic activity.

2. Diffusion related problems seem to reflect: the chemical affinity of the substrate for the support; the rate of diffusion of the substrate into the support. Systematic kinetic studies as described above should be applied to new supports with a wide range of polar groups. These supports should provide a wide range of affinities and allow careful matching of substrates to supports. Selectivities due to differences in diffusion rates can be modified by solvent effects.

3. Kinetic studies should try to correlate activity and selec-

tivity to the nature of the polymeric structure (distance between the catalytic species, bifunctionality, tacticity, etc.).

4. Very precise kinetic studies should compare the rate of a simple reaction in the presence of a solvent with that in its absence (liquid phase reaction vs. gas phase reaction).

5. Stabilized coordinative unsaturation, a characteristic of heterogeneous catalysis, should also occur with supported complexes. Measurement of the reactivity of coordinatively unsaturated species could bring some insight into pure homogeneous catalysis.

6. Kinetic studies on multifunctional systems prepared by proximal linking of two different catalysts would determine whether or not increased rates can be obtained in multistep reactions.

c. Catalytic Systems. Hydrogenation and hydroformylation of olefins have received the most study to date. Emphasis should be on:

1. asymmetric synthesis (recovery of the costly chiral ligand, use of natural chiral polymers as ligand),

2. production of new catalysts without homogeneous counterpart,

3. oxidation catalysts,

4. bimetallic systems resulting from proximal binding (which might, for example, be capable of simultaneous nitrogen and oxygen activation on adjacent sites),

5. multifunctional catalysts (which might be effective in the catalysis of processes involving thermodynamically unfavorable intermediates, e.g. B in a reaction of the type $A \rightleftharpoons B \rightarrow C$),

6. Any homogeneous catalyst whose deactivation is a result of dimerization,

7. the support of molecular clusters (whether or not their exact solution geometry is maintained).

d. Characterization. Characterization of supported complexes has been very poor (one usually refers to "closely related soluble counterpart"). No definite proof of any structure of a supported complex has been reported. Local concentrations reported are only the result of statistical estimates of ligand distribution. Significant progress is impossible unless new techniques are exploited:

1. high resolution NMR spectroscopy (for polymers with little crosslinking),

2. broad-band NMR spectroscopy (solid state),

3. UV spectroscopy,

4. ESCA,

5. ESR,

6. Extended X-ray absorption fine structure,

7. High-resolution electron microscopy, electron microprobe spectroscopy, small-angle X-ray scattering (supported clusters).

Characterization would also be greatly aided by better definition of the physical and chemical properties of supports.

  e. Linking Groups. Desirable properties for linking ligands are
chemical inertness and non-lability. Phosphines grafted to organic
polymers and inorganic supports have been most studied. However, they
are easily oxidized and labile ligands which limit the types of reactions which can be carried out on the supports. Better linking agents
are needed to meet the requirements listed. There should be further
development of non-labile ligands such as macrocyclic amines, chelating phosphines, and carboranes.

  New methods for functionalizing polymers and inorganic oxides
should also be developed. Particular attention should be paid to
those which would result in a more controlled distribution of linking
ligands along the polymer backbone or on a surface (i.e., copolymerization of prefunctionalized monomers).

  2. The Molecular Nature of the Complex is Not Preserved

  By this approach a number of high-performance propylene polymerization catalysts have recently been developed in industry (ICI,
Montedison, Solvay). The general method consists of direct reaction
of an organometallic with the surface group of an inorganic support,
followed by treatment with a cocatalyst (a Lewis acid or reducing
agent). Oxides are frequently used as supports.

  The resulting catalysts are much more reactive than unsupported
systems. Unfortunately, in this case where the molecular character
of the catalyst is not preserved, there is a complete lack of information on the chemical nature of the resulting catalyst. Moreover,
the preparation is very empirical and does not proceed from knowledge
of the nature of the interaction between support and organometallic
species.

  To overcome these limitations and to extend the field to other
catalytic reactions, some studies should be carried out: one should
search for reactions of organometallics (carbonyls, nitrosyls, cyclo-
pentadienyls, σ-alkyls, σ-allyls, metal clusters) with the functional
groups present on the surface of inorganic oxides (e.g., Si-OH, Al-OH,
-Al-, Al-O-Al, Si-O-Si), with particular emphasis on characterization
of the resulting grafted organometallics.

  Kinetic studies should be carried out to assess the dependence of
reactivity upon differences in various properties of supports.

  There should be synthetic studies of any soluble compounds which
model the active surface species. The organometallic chemistry of
silicon and aluminum derivatives should be developed in this direction.

  A particularly important class of organometallics which should be used as starting materials in this approach is that of metal clusters. They can serve as precursors for supported metal particles, and these particles may have precisely defined nuclearity and composition otherwise unachievable. It may thus be possible to increase our knowledge of supported small metal particles. This research area in particular would benefit from close interaction between workers in homogeneous and in heterogeneous catalysis.

3. Related Areas

New fields of research which offer new approaches to catalyst recovery and re-use:

a. Membrane systems which are permeable only to reactants and products but not to catalysts.

b. New ligands which allow catalysis in biphasic (liquid, e.g., water) systems. Also included in this area are bulky ligands to be used in membrane systems.

c. Supported enzymes, which while out of the scope of this report, should provide a source of new supports and new grafting techniques available to supported complexes.

B. METAL CLUSTERS IN CATALYSIS

It is now possible to prepare molecular clusters containing as many as 18 transition metal atoms. They contain a central core of metal atoms, bonded together in many different geometries, surrounded by ligands. The size of these molecules can in fact approach that of particles of very finely dispersed metals, and some large clusters have structures resembling those proposed for metal crystallites. It has thus been suggested that the chemistry of clusters and of small metal particles may have features in common.

However, in clusters we have very small metal frameworks stabilized by appropriate ligands, whereas in crystallites they are probably stabilized by interaction with an oxide surface, or by the framework of the zeolite or other matrix in which they are trapped. Determination of the precise relationship between the two systems is therefore important.

This determination will require the development of new physical methods to establish size and geometry -- primarily for use with crystallites, as molecular clusters in the solid state can be well-characterized by X-ray diffraction. Some of the most promising new techniques for crystallites are high-resolution electron microscopy, Auger and ESCA spectroscopy, and Extended X-ray Absorption Fine Structure. Only where the size and geometry of metal crystallites have been unambiguously established will it be possible to determine the degree of their correlation with molecular metal clusters.

It is also obviously important to look more closely at the
effect of the surrounding medium on the stability of a particular
cluster framework, both in crystallites (Very Small Particles) and
in molecular clusters (MC). Equilibria of the following sort should
be examined when possible:

$$VSP \underset{-L}{\overset{+L}{\rightleftharpoons}} MC$$

Molecular metal clusters will most nearly resemble metal crystallites
when the latter have full chemisorption on the available metal surface. Thus chemical and spectroscopic information on the reactivity
and bonding of cluster ligands (e.g., H_2, O_2, CO, olefins) may give
information on the properties of chemisorbed species on such fully
covered surfaces. Similarly, the chemical physics and organometallic
chemistry of metal surfaces must be expanded in order to obtain un-
ambiguous characterization of the corresponding surface species.

A final important question is the following: Is there a gap of
properties, or a kind of progressive change, from the molecular state
to a multicenter metallic framework? Theoretical calculations of high
quality are of interest on this point. It must be emphasized that
the relationship between molecular clusters and chemisorbed very small
crystallites is qualitative and presently supported by very little
direct experimental evidence.

Quite aside from their possible resemblance to metal surfaces
and crystallites, dissolved metal clusters show considerable promise
in homogeneous catalysis. Given the rudimentary level of our knowledge in this area, the most practical way of discovering such applications is empirical: there should be widespread investigation of the
reactions of many clusters with many substrates. Increased experimental work is desirable on all stoichiometric and catalytic reactions
of clusters.

For the systematic study of catalytic reactions clusters should
be readily available, and attention should be paid to practical large-
scale preparative methods. Here a more systematic approach is possible and desirable. The success of the redox condensation approach
in preparing polynuclear carbonyl anions suggests that other general
reactions should be sought for cluster synthesis. Electrochemical
methods and metal atom techniques (which may also be useful for preparing nonligated crystallites) should be applied in this area.

Difficulty in analyzing the results of cluster synthesis reactions has resulted from the fact that definitive characterization
is available only from X-ray crystallography. Other physical methods,

especially those applicable in solution, need to be developed. Possibilities include nmr (of nuclei other than ^{1}H and ^{13}C), cyclic voltammetry, Raman spectroscopy (newly useful because of technological advances), and electronic spectra. Finally, the well established usefulness of mass spectrometry for the characterization of isolated clusters would be increased by the wider availability of field absorption instrumentation.

Metal clusters may offer a number of advantages over mononuclear systems as homogeneous catalysts. Some of these advantages have been established, others have been claimed, and all merit further investigation:

1. Substrates can be activated by binding to several metal centers at once, e.g., ethylidyne tricobalt systems.
2. Several different substrates can be bound on adjacent metal atoms, resulting in precise control of reactant geometries and lower entropies of activation.
3. Many clusters (particularly the ones containing metal halides) are capable of delivering or receiving several electrons without major changes in geometry, and they thus have a low activation energy for multiple-electron oxidation and reductions.
4. Coordination of a substrate center may strongly influence the reactivity of neighboring metal atoms, leading to the possibility of cooperative effects and enhanced reaction rates.

Proof that these advantages are realized in catalytic practice will be difficult to obtain. Indeed, the demonstration that an active species is in fact polynuclear itself poses a major challenge to mechanistic chemists. Detailed fundamental work is required on the mechanisms of even the simplest reactions. The most useful tools are likely to be kinetics and the isolation of model compounds for intermediates.

The mobility of the ligands present in a cluster is of interest both as compared with that on surfaces and as an index of simultaneous availability of substrates at a given spot on the cluster framework. Mobility of metal atoms is also of possible significance for the changes in composition associated with diffusion at metal surfaces (aging of catalysts). Furthermore, measurement of metal atom mobility will allow investigation of the rigidity of cluster frameworks, and comparison of the results for clusters with those for crystallites.

Metal clusters present a unique opportunity for the synthesis and the study of interstitial compounds which may be formed during industrial catalytic reactions. Their relevance there and in metallurgy makes desirable an exhaustive study of interstitial carbides and hydrides and the search for other similar species such as nitrites.

Another important area for future research is that of metal

frameworks containing different metal atoms, particularly from different transition periods (readily identified by X-ray analysis). The
electronic and steric properties of these mixed clusters should be
investigated, as well as the differences in reactivity between metals
in mixed clusters and homonuclear clusters. Mixed clusters are also
of interest for the preparation of bimetallic or trimetallic supported
catalysts, which will be more precisely defined than those available
by present techniques.

Finally, we recommend the study of the reactivity and catalytic
properties of "naked clusters" such as $[Pb_5]^{-2}$, $[Sn_9]^{-2}$, and $[Bi_9]^{+5}$.
These metal frameworks could coordinate reactant molecules in the
same way that a heterogeneous catalyst metal surface does, i.e., without the additional stabilizing ligands found on conventional metal
clusters.

C. PHASE-TRANSFER TECHNIQUES IN HOMOGENEOUS CATALYSIS

There is a need for new techniques for the preparation of organometallic catalysts in situ: quite often they are currently prepared
by reduction with expensive and dangerous reagents such as aluminum
alkyls. The phase-transfer technique provides a way to introduce
ionic nucleophiles as CN^- and NCO^-, and particularly reducing ones
such as BH_4^- and PhO^-, easily into the organic phase.

In this way homogeneous catalysts may easily be generated, e.g.,

$$Pd(acac)_2 \quad + \quad PPh_3 \quad \xrightarrow[NR_4X]{PhOH/Bz} \quad Pd(PPh_3)_2(OPh)_2$$

$$Pd(PPh_3)_2(OPh)_2 \quad + \quad NaBH_4 \quad \xrightarrow[NR_4X]{PPh_3/Bz} \quad Pd(PPh_3)_4$$

Another important application involves homogeneous catalytic
processes in which nucleophiles are involved as substrates, e.g.,

$$RX \quad + \quad CO \quad + \quad NaOH \quad \xrightarrow[NaOH/Bz]{Pd(PPh_3)_4} \quad NR_4X \quad + \quad RCOONa \quad + \quad NaX$$

This technique can afford several advantages:

1. Better activity (higher turnover numbers for the homogeneous
catalyst).
2. Easy activation of hydrocarbons with active hydrogen to

produce nucleophiles involved in catalytic processes (e.g.,
Ph-C≡C-H → Ph-C≡C⁻).
    3. Use of NaOH in place of reagents as $NaOCH_3$, NaH, $NaNH_2$.
    4. Easy recovery of catalyst uncontaminated by inorganic salts.

    Very recently rather complex phase-transfer catalysts (e.g.,
crown ether cations) have been used as salts of a polymeric anionic
matrix. In this way a triphase system has been produced, permitting
catalyst recycling by simple filtration from the reaction medium.
    The applications of this technique to organometallic chemistry
and homogeneous catalysis are at a very preliminary stage. A large
amount of experimental work is recommended in order to determine the
most likely areas for successful applications.

D.   SUMMARY OF RECOMMENDATIONS

    1. Better characterization techniques for both clusters in
solution and for supported complexes.
    2. Tailor-made supports for specific reaction catalysts.
    3. Quantitative study of diffusion rates in supports and their
influence on reaction rates.
    4. Development of new systems which fully exploit the advantages
of supports, such as stabilized, highly unsaturated catalysts and
multifunctional catalysts.
    5. Use of clusters to generate small metal particles of con-
trolled nuclearity and composition.
    6. Systematic syntheses of metal clusters.
    7. Quantitative comparison of clusters and small metal crystal-
lites.
    8. Study of the mechanisms of reactions of metal clusters.
    9. Measurement of the rigidity of the framework of metal clus-
ters.

Bibliography for Clusters

E. L. Muetterties, Bull. Soc. Chim. Belg., 84, 959 (1975).
R. Ugo, Catal. Rev., 11, 225 (1975).
J. M. Basset and R. Ugo, in Aspects of Homogeneous Catalysis, Vol.
III, R. Ugo, ed., Reidel (Holland, (1976).
B. R. Penfold, Perspect. Struct. Chem., 2, 71 (1968).
P. Chini, G. Longoni, and V. G. Albano, Adv. Organometal. Chem., 14,
285 (1976).
J. Lewis and B. F. G. Johnson, Pure Appl. Chem., 44, 43 (1975).

V. CATALYTIC PROCESSES

          Co-Chairmen: G. Pez, A. Sacco
          Members: J. E. Bercaw, H. H. Brintzinger, G. P. Chiusoli,
                    H. Felkin, G. M. Giongo, A. Langer, G. Mestroni,
                    T. J. Meyer, G. Paiaro, A. Panunzi, P. Pino,
                    A. Umani-Ronchi

A. Metal Complex Catalyzed Introduction and Transformation of
                        Functional Groups

     The use of CO and O_2 for such chemistry has been discussed else-
where in the Report.

     The vast majority of steps in the synthesis of organic compounds
involve the introduction and transformation of functional groups by
conventional processes. Many of these processes are laborious and
wasteful of energy, and many steps are often necessary to effect selective transformations. It would therefore be highly desirable to
replace some of these processes by more direct and more selective
catalytic ones. In a few cases, this has already been done. For
example, adiponitrile is now made directly from butadiene and HCN by
a catalytic process involving a nickel complex. Another advantage of
the introduction of catalytic processes might be the elimination of
some dangerous or highly polluting reagents $(CH_3)_2SO_4$, $COCl_2$, etc.
and solvents (THF, HMPA, CCl_4, etc.).

     Obviously it is impossible in the foreword of this report to
cite all the reactions which could be improved by the application of
metal complex catalysts. We restrict ourselves to the following
(rather arbitrarily chosen) examples.

1. Introduction of Functional Groups. a) Olefins and acetylenes

     - the direct addition of NH_3 onto ethylene and acetylene to yield
amines and N-heterocycles.
     - the addition of H_2O to olefins to make primary alcohols, al-
dehydes and ketones.

225

- various addition reactions of CO_2, SO_2, sulfur, RCOOH, and HCN onto olefins and acetylenes.
- use of nitriles or HCN with acetylenes for the formation of N-heterocycles.

b) Aromatics
- regioselective introduction of functional groups, e.g., $-NO_2$, $-COOH$, $-SO_3H$, $-Cl$, $-CN$ into aromatic carbo- and hetero-cycles.
- direct functionalization of saturated (sp3) carbon atoms. In some cases, e.g. with steroids, it will be necessary to devise regio- and stereospecific methods for this.

2. Transformation of Functional Groups
- hydrogenation of functional groups containing C-O and C-N bonds such as the conversion of -COOH to $-CH_2OH$, or C=O to CH_2, or $-C-NO_2$ to $-C-NH_2$.
- carbon-carbon bond formation from substrates containing carbonyl groups by alkylation and acylation procedures.
- the Friedel-Crafts reaction.
- the Beckmann or Fries rearrangements, etc.
- hydrolysis of nitriles, hindered esters, amides, under mild conditions.
- formation of carboxylic esters and amides, under mild conditions.
- replacement of $-SO_3H$, $-NO_2$, etc. on aromatics with other functionalities.
-dealkylation of $-NR_2$ or $-OR$ groups.

In this very wide range of chemistry there are a certain number of basic transformations between the reacting chemicals and a metal catalyst. In particular we need to develop our knowledge of the following aspects of metal-complex catalysis:

1. The formation of metal-hydrogen bonds and their reactivity. It should be particularly valuable to quantify the highly variable acidity of transition-metal hydrides.
2. The formation of metal-carbon bonds, e.g., by the addition of hydrides to olefins and acetylenes, by de-halogenation or de-oxygenation of halides or oxygenated compounds, as well as by carbon-carbon bond breakage (possibly via carbenes).
3. The latter two points imply a study of ligand substitutions and ligand reactions.
4. The conditions for the attachment of substrate to the metal (e.g., coordination).
5. The reaction of an external reagent with the coordinated substrate. This includes a study of insertion reactions and of nucleophilic and electrophilic attack on the substrate. The ligand environment can markedly affect this process and its influence should be studied; this may imply the design of new ligands.
6. Ion pair effects. This may be important in directing in-

sertion reactions with ionic metal complexes.

7. Redox systems, especially for cases in which reagents are oxidatively added on to organic substrates.

8. Elimination of metal in metal-induced reactions so as to render these catalytic.

B. Asymmetric Synthesis by Homogeneous Organometallic Catalysts

1. Importance of this Research Field. Prochiral substances of the general type $R_1R_2C=X$ where $X=CR_2$, NR, or O are among the most available and most useful starting materials for chemical syntheses. Many transformations of these prochiral compounds, such as hydrogenations, hydrosilylations, dimerizations, polymerizations, and various functionalization reactions (hydration; hydrocyanation; methoxycarbonylation and others) will in general lead to asymmetric products by enantioface discrimination. Enantioelective reactions (kinetic resolution) by chiral catalysts might equally well lead to optically active products from racemic precursors. In many instances it is essential for the ultimate use of these materials that the products be obtained as pure enantiomers, particularly for their application as pharmaceuticals or as food additives for the remedy of nutritional deficiencies. While in some cases enzymatic processes are available and preferable for these asymmetric transformations, this is not always the case. The production of either natural or non-natural amino-acid enantiomers is an example for the superior flexibility of homogeneous organometallic catalysts compared to enzymatic processes. A related demand is that for racemization catalysts for the conversion of readily available chiral materials to their more useful epimers.

In addition, studies of asymmetric reactions will in many instances constitute a powerful tool for the elucidation of catalytic reaction mechanisms in general.

2. State of the Art. Of the studies published, the majority is concerned with three types of reactions: hydrogenation, hydrosilylation and hydroformylation. Other reactions, such as α-olefin oligomerization, have been the subject of a limited number of publications. In olefin hydrogenation, high optical purities, approaching those of enzymatic reactions, have been obtained, but only for functionalized molecules, where multiple interaction appears to assist the selection of prochiral R- or S-substrate approach. Metal ligand systems employed for these reactions have been largely based on chiral phosphine derivatives. Relatively few studies have been published on other chiral ligand systems such as amides, amines, sulfoxides, and others. Among the metals studied, Rh is by far the most prominent, other metals such as Pd, Ni, or Co have been used in a few instances only for the reactions named above.

   3. Desirable Developments. A main objective for the development
of this field at the present time would be to derive more generally
applicable rules and principles concerning the transfer of chirality
by a catalyst. To this end, a broader range of chiral ligand and metal structures including natural and synthetic chiral polymers should
be investigated for chiral catalytic activity with respect to an
expanded spectrum of asymmetric transformations including enantioselective catalytic oxidation of racemic substrates.

   Equally important, would be the description of these reaction
systems not only in terms of kinetic data, but also in terms of
structural information on catalysts and catalyst-substrate complexes
in solution (for instance, from careful spectral studies). Of primary interest is, of course, the state of static and dynamic chirality of the metal-ligand system by itself, e.g., stereochemical rigidity and its relation to catalytic stereoselectivity. More important,
but also more difficult to obtain, are studies on substrate-catalyst
interactions. Some information of this type might come from interaction of the chiral complex with a prochiral substrate under non-
catalytic conditions (i.e., at low temperatures, in the absence of
co-substrates, etc.). The ultimate aim of such a study would be the
identification of the structure of crucial diastereoisomeric pairs and
of the type and strength of repulsive and attractive interactions
which govern the stereochemical outcome of the catalytic reaction.

   Considerable gains in this regard can be expected from a systematic reinvestigation of known processes and from a consistent explanation of observations concerning the dependence of chiral selectivity on substrate structures and on the location of the asymmetric
center relative to the metal-substrate reaction center for differing
types of reactions and of effects such as those of varying counter-
ions on the stereoselectivity of cationic complexes.

   One can reasonably hope that working hypotheses (similar to
those applicable, to stoichiometric reactions with chiral Li-, B- and
Al-hydride and alkyl coordination compounds) might become available
to guide further efforts concerning the synthesis and screening of
stereospecific catalysts. An example of such a working hypothesis,
yet to be tested, is the one proposed by Felkin: that optimal stereoselectivity in a related class of catalysts is associated with the
presence in the metal-ligand structure (i.e., before association with
the prochiral substrate) or a C_2 axis.

                   C. Photocatalytic Processes

   Light energy can be used to influence the course of homogeneous
catalytic reactions in two ways. Spontaneous or exoergic chemical
reactions may be accelerated by irradiation (photo-assisted catalysis). The increase in reaction rate occurs by the opening of one or

several new kinetic paths by photochemical changes in the catalyst system. In the second case, light is used as a source of chemical potential, to drive an endoergic reaction to completion. This type of process may be called chemical photosynthesis.

1. Photoassisted Catalysis. Irradiation of a metal complex catalyst may have the following effects: (a) result in the expulsion of ligands to yield coordinatively unsaturated sites; (b) change the ligand environment; (c) alter the redox properties of the metal center. The reactivity patterns of the metal might thus be sufficiently altered to cause favorable changes in the performance of the catalyst. The excitation is essentially a localized effect at the catalytic site and this offers a number of advantages (i.e., ease of application, containment, use of milder conditions, etc.) over the usual thermal systems. Certain requirements for photoexcitation must be met, notably: (a) catalyst needs to be designed for efficient use of impinging radiation; (b) the catalyst must not be photochemically degradable.

2. Chemical Photosynthesis. Some metallic oxides (e.g., TiO_2) are known to function as photocatalysts. However, to the best of our knowledge, no homogeneously catalyzed endoergic photolytic systems are known at present. In the past five years, the elementary kinetic and energetic steps for one potentially useful photochemical system have been clearly established. The example is the long lived excited state of tris(2,2'-bipyridyl)ruthenium(II).

From estimated thermodynamic properties of $(bipy)_3Ru(II)^*$, we note that it can function either as an oxidizing agent (to convert H_2O to $1/2O_2$) or as a reducing agent (to convert H_2O to H_2). However, its use as a photocatalyst (e.g., for the splitting of H_2O into H_2 and O_2) has yet to be realized.

Conclusions and Recommendations

Photoactivated systems potentially can have a very wide range of applications ranging from the improvement of present day catalytic processes (photoassisted catalysis) to the synthesis of basic chemicals from natural raw materials (CH_3OH from CO and H_2O) as well as for the generation of fuels. For the endoergic reactions, sunlight appears to be the only practical source of energy, which for its economical utilization requires the use of very inexpensive catalytic systems.

We recommend that further fundamental work be carried out on the photochemistry of metal complexes; detailed suggestions in this area have appeared in a previous NSF report (Inorganic Photochemistry Related to Transfer, Storage Conservation, and Conversion of Energy, June, 1975). On a broader front it is obviously desirable to promote coupled research efforts in the combined areas of photochemistry and homogeneous catalysis. Only in this manner will catalytic chemistry, based on the use of molecular excited states eventually become practicable.

D. N_2 Activation for its Eventual Use in Homogeneous Catalysis

N_2 is a very abundant raw material for the chemical industry, but is so inert that considerable energy is required for its processing. The catalytic reduction of N_2 to NH_3 at high temperatures and pressures has been carried out with a heterogeneous catalyst for over fifty years; yet no efficient non-enzymatic homogeneous catalyst for N_2 fixation is now known. Despite the fact that the major expenditures in ammonia synthesis are in the preparation of the N_2/H_2 synthesis gas, the development of a homogeneous catalyst for the conversion of N_2 and H_2 to NH_3 under mild conditions could offer some energy savings in production and transportation. Of greater potential significance would be the specific, direct conversion of N_2/H_2O and energy to N_2H_4, since this would open the way to the utilization of hydrazine as a fuel, for example, in electrochemical cells; such specificity in N_2 reduction may be possible only with homogeneous catalysts.

At present the synthesis of all industrial organic nitrogen compounds ultimately rely on NH_3 from the Haber/Bosch process. Thus the development of homogeneous catalysts utilizing N_2 as a substrate for the direct synthesis of amines and nitrogen heterocycles would be of great significance. Furthermore, homogeneous catalysts which promote the oxidative fixation of N_2 to nitric acid or nitrates would be highly desirable.

No efficient homogeneous catalyst for N_2 fixation is now known, and only a few systems have been found in which a soluble transition metal complex promotes the stoichiometric reduction of N_2 to N_2H_4 or NH_3. Many transition metal dinitrogen complexes have been isolated, but N_2 activation has been demonstrated thus far in only two types of compounds: $[ML_4(N_2)_2]$, (M = Mo, W; L = PR_3) and $Cp_2MN_2·N_2$, (M = Ti, Zr; Cp = cyclopentadienyl) and here the mechanism of N_2 reduction is only partially understood. The former class of compounds have been shown to undergo facile, stoichiometric reactions with organic halides to generate alkylated N_2 moieties, but few other dinitrogen complexes have been investigated in terms of their reactivity other than with respect to reduction to ammonia and hydrazine. Thus the chemistry for modifying N_2 coordinated to transition metals is still largely unknown.

Recommended Studies

The development of homogeneous catalytic systems utilizing N_2 as a substrate will require a better understanding of the chemical reactivity of coordinated dinitrogen. Fundamental studies are recommended in the following areas:

1. Synthesis of transition metal dinitrogen complexes with reactive N_2, especially those with co-ligands other than tertiary phosphines.

2. Definition of the reactive intermediates and mechanism of reduction in N_2 reducing systems.

3. Investigations of transition metal promoted reactions of N_2 with organic substrates (RX, alkenes, acetylenes, etc.).

4. Initiation of a search for soluble (or insoluble) transition metal compounds capable of promoting the oxidation of N_2 to HNO_3 or NO.

5. Fundamental studies of the coordination chemistry of ligands representative of suspected intermediates in N_2 reduction (e.g., $-N=NR$, $RN=NR$, $-NRNR_2$, $=NR$, $\equiv N$) and oxidation (e.g., NNO, N_2O_2, etc.).

6. Attempts to develop stoichiometric N_2 fixing systems into catalytic systems for example, by electrochemical means. In this regard, it would be desirable to initiate fundamental studies of electron transfer to coordinated N_2 in transition metal dinitrogen complexes.

LIST OF CONTRIBUTORS

BELGIUM:

Prof. Ph. Teyssié
Laboratoire de Chemie
Macromoleculaire et de,
Catalyse Organique
Université de Liegè, Sart-Tilman
4000 Liegè

CANADA:

Dr. Brian James
Department of Chemistry
University of British Columbia
Vancover 8, British Columbia

FRANCE:

Dr. J. M. Basset
Institut de Reserches sur la
        Catalyse
79, Boulevard du 11 Novembre 1918
69626 Villeurbanne-CEDEX

Prof. H. Felkin
Institut de Chemie des Substances
        Naturelles
CNSR, 91190- Gif sur Yvette

Dr. H. Mimoun
Institut Francais du Pétrole
Division Recherches
Chimiques de Base
1 et 4 Avenue de Bois-Préau
Boite Postal 18
92502 Rueil-Malmaison

Prof. J. Osborn
Institut de Chemie
1, rue Blaise Pascal
Universite "Louis Pasteour"
67000 Strasburgo

Dr. I. Tkatchenko
Institut de Recherches sur la
        Catalyst
79, Boulevard du 11 Novembre 1918
69626 Villeurbanne-CEDEX

WEST GERMANY:

Prof. H. Britzinger
Fachbereich Chemie
Universitat Konstanz
D-775 Konstanz, Postfach 733

HOLLAND:

Dr. C. Masters
Koninklijke Shell Laboratorium
Shell Research B.V., Postadres
Postbus 3003, Amsterdam

Dr. R. Sheldon
Koninklijke Shell Laboratorium
Shell Research B.V., Postadres
Postbus 3003, Amsterdam

JAPAN:

Prof. Yoshio Ishii
Department of Synthetic Chemistry
Faculty of Engineering
Nagoya University, Chikusa
Nagoya

Prof. Jiro Tsuji
Tokyo Kogyo Daigaku, Ookayama
Meguro-ku, Tokyo

Prof. Akio Yamamoto
Tokyo Kogyo Diagaku, Ookayama
Meguro-ku, Tokyo

NORWAY:
Dr. I. E. Ruyter
NIOM, Forskningsvn 1, Olso

SWITZERLAND:

Prof. P. Pino
Eidgenossische Techische
Hochschule Zurich
Technisch-Chemisches
        Laboritorium
Universitatstrasse 6, CH-8006
Zurich

ITALY:

Prof. U. Belluco
Instituto`di Chimica Industriale
della Facolta di Ingegeria
Via Marzolo, 21, Padova

Prof. G. Braca
Instituto di Chimica Organica
Industriale dell'Università
Via Risorgimento, 35, Pisa

Prof. G. Cainelli
Instituto Chemico CIAMICAN
Via A. Selmi 2, Bologna

Prof. B. Calcagno
SIR S.P.A., Via Grazoili, 33
Milano

Prof. F Calderazzo
Istituto di Chimica Generale
e Inorganica dell'Università
Via Risorgimento, 35, Pisa

Dott. L. Cassar
Istituto Donegani,
MONEDISON - DIRS, Novara

Prof. S. Cenini
Istituto di Chimica Generale
e Inorganic dell'Università
Via G. Venezian, 21, Milano

Prof. P. Chini
Instituto di Chemical Generale
e Inorganica dell'Università
Via G. Venezian, 21, Milano

Dott. M. Clerici
SNAM Progetti, Laboratori Riuniti
San Donato Milanese, Milano

Prof. F. Conti
MONTEDISON - DIPE RT
Largo Donegani 2, Milano

Prof G. Costa
Istituto di Chimica Fisica
dell'Università, Trieste

Prof. G. Dolcetti
Istituto di Chemica General
a Inorganica dell'Università
della Calabria, Cosenza

Dott. G. M. Giongo
SNAM Progetti, C.P. 15,
00015 Monterotondo - Roma

Prof. N. Giordano
Istituto di Chemica Industriale
dell'Università, 64, Messina

Prof. D. Guisto
Università di Milano
Institute di Chemica Generale
20133 Milano-via G. Venezian,21

Prof. M. Grazinani
Cattedra di Chimica Generale
e Inorganica dell'Univerità,
Trieste

Dott. R. Leoni
SNIA Viscosa S.P.A., Rierca di Base
20031 Cesano Maderno, Milano

Dott. A. Ligorati
SIR S.P.A., Via Grazioli, 33,
Milano

Prof. G. Mestroni
Istituto de Chimica
Universita di Trieste, Trieste

Prof. G. Modena
Istituto di Chimica Organica
dell'Univerita
Via Marzolo, 1, Padova

Prof. G. Paiaro
Istituto di Chemica
Analitica dell'Universita
Via Loredan, 4, Padova

Prof. A. Panunzi
Istituto di Chimica Generale
dell'Universita, Via Mezzocannone
Napoli

Prof. Piacenti
Istituto di Chimica Organica
dell'Università
Cattedre di Chimica Organica
Industriale, Firenze

Prof. M. Rossi
Istituto di Chimica Generale
e Inorganica dell'Università
Via Amendola, 173, Bari

Dott. P. Rossi
SNIA Viscosa S.P.A.
Ricerca di Base, 20031
Cesano Maderno, Milano

Prof. A. Sacco
Istituto di Chimica Generale
e Inorganica dell'Università
Via Amendola, 173, Bari

Prof. G. Sbrana
Istituto di Chimica Organica
Industriale
Via Risorgimento, 35, Pisa

Prof. R. Ugo
Università di Milano
Institute di Chimica Generale
20133 Milano-via G. Venezian,21

Prof. A. Umani-Ronchi
Istituto Chimico CIAMICIAN
Via A, Selmi, 2, Bologna

UNITED STATES:

Dr. James E. Bercaw
Div. of Chemistry and Chemical
Engineering
California Institute of Technology
1201 E. California Blvd.
Pasedena, California 91109

Dr. Charles P. Casey
Department of Chemistry
University of Wisconsin
Madison, Wisconsin 53706

Dr. D. Robert Coulson
E.I. DuPont de Numours & Co.
Wilmington, Delaware 19899

Dr. Darrel Fahey
Phillips Petroleum Co.
Research and Development Division
Bartlesville, Oklahoma 74004

Dr. Denis Forster
Monsant Co.
800 N. Lindbergh Blvd.
St. Louis, Missouri 63066

Dr. Robert H. Grubbs
Department of Chemistry
Michigan State University
East Lansing, Michigan 48824

Dr. Jack Halpern
Department of Chemistry
University of Chicago
Chicago, Illinois 60637

James E. Lyons
Suntech, Inc.
Research and Engineering Division
P.O. Box 1135
Marcus Hook, PA 19061

Arthur W. Langer, Jr.
Exxon Research and Engineering Co.
P.O. Box 45
Linden, New Jersey 07036

Dr. Thomas J. Meyer
Department of Chemistry
University of North Carolina
at Chapell Hill
Chapell Hill, North Carolina
27514

Dr. Jack R. Norton
Department of Chemistry
Princeton, University
Princeton, New Jersey 08540

Dr. Guido P. Pez
Chemical Resaerch Center
Allied Chemical Corporation
P.O. Box 1201R
Morrison, New Jersey   07960

Dr. Richard B. Schrock
Department of Chemistry
Massachusetts Institute of Technology
77 Massachusetts Ave.
Cambridge, Massachusetts   02139

Dr. Karl B. Shapless
Department of Chemistry
Massachusetts Institute of Technology
77 Massachusetts Ave.
Cambridge, Massachusetts   02139

Dr. John K. Stille
Department of Chemistry
Colorado State Univerity
Fort Collins, Colorado   80521

Dr. Minoru Tsutsui
Department of Chemistry
Texas A & M University
College Station, Texas   77843

Dr. Oren Williams
National Science Foundation
Washington, D.C.   20550

SUBJECT INDEX